T0313322

Recent Advances in Information, Communications and Signal Processing

RIVER PUBLISHERS SERIES IN SIGNAL, IMAGE AND SPEECH PROCESSING

Series Editors

MONCEF GABBOUJ
Tampere University of Technology
Finland

THANOS STOURAITIS
University of Patras
Greece

Indexing: All books published in this series are submitted to the Web of Science Book Citation Index (BkCI), to CrossRef and to Google Scholar.

The "River Publishers Series in Signal, Image and Speech Processing" is a series of comprehensive academic and professional books which focus on all aspects of the theory and practice of signal processing. Books published in the series include research monographs, edited volumes, handbooks and textbooks. The books provide professionals, researchers, educators, and advanced students in the field with an invaluable insight into the latest research and developments.

Topics covered in the series include, but are by no means restricted to the following:

- Signal Processing Systems
- Digital Signal Processing
- Image Processing
- Signal Theory
- Stochastic Processes
- Detection and Estimation
- Pattern Recognition
- Optical Signal Processing
- Multi-dimensional Signal Processing
- Communication Signal Processing
- Biomedical Signal Processing
- Acoustic and Vibration Signal Processing
- Data Processing
- Remote Sensing
- Signal Processing Technology
- Speech Processing
- Radar Signal Processing

For a list of other books in this series, visit www.riverpublishers.com

Recent Advances in Information, Communications and Signal Processing

Editors

Andy W. H. Khong

Yong Liang Guan

Nanyang Technological University
Singapore

LONDON AND NEW YORK

Published 2018 by River Publishers
River Publishers
Alsbjergvej 10, 9260 Gistrup, Denmark
www.riverpublishers.com

Distributed exclusively by Routledge
4 Park Square, Milton Park, Abingdon, Oxon OX14 4RN
605 Third Avenue, New York, NY 10017, USA

Recent Advances in Information, Communications and Signal Processing / by
Andy W. H. Khong, Yong Liang Guan.

Routledge is an imprint of the Taylor & Francis Group, an informa
business

ISBN 978-87-93609-43-3 (print)

While every effort is made to provide dependable information, the
publisher, authors, and editors cannot be held responsible for any errors
or omissions.

Contents

5   Location Template Matching on Rigid Surfaces for Human–Computer Touch Interface Applications   137

Nguyen Q. Hanh, V. G. Reju and Andy W. H. Khong

6   Automatic Placental Maturity Grading via Deep Convolutional Networks   165

Baiying Lei, Feng Jiang, Yuan Yao, Wanjun Li, Siping Chen, Dong Ni and Tianfu Wang

Preface

This book contains some of the latest development in the fields of Information, Communications and Signal Processing. Each chapter reviews recent challenges that have been identified by fellow researchers and describe new algorithms that are proposed to tackle the associated challenges.

This book covers two broad areas – communications and signal processing. The first two chapters discuss two extremely hot topics in the 5th-generation cellular mobile technology, namely, NOMA (non-orthogonal multiple access) and millimeter-wave radio. Chapter 3 focuses on multi-session network coding, a relatively less visited area of the popular network coding literature. Chapter 4 is dedicated to a practical aspect of the capacity-approaching LDPC (low density parity check) code: iterative scheduling of LDPC decoding. With the advent of machine learning techniques, a chapter has been dedicated to discuss challenges associated with placental maturity grading via medical imaging. Another chapter describes the use of vibration signals and source localization algorithms to convert an ordinary surface into a touch interface.

This book serves as a guide for researchers and graduate students who wish to understand challenges associated with each topic area and to implement some of the well-known algorithms that have been developed. This book is also suitable for researchers and engineers in the industry who are keen to bring new technologies to the market. Readers are encouraged to contact the authors to establish collaboration opportunities.

Acknowledgements

This book is dedicated to fellow researchers and learners who inspired us through their relentless pursuit of knowledge and constantly applying knowledge gained to solve real-world problems. Thanks to all authors who have shown tremendous support and commitment for their contributions to this book. On behalf of the authors, the editors wish to thank River Publishers for their effort in ensuring the successful completion of this book.

List of Contributors

A. Anwar, *Department of Electrical and Electronic Engineering, Auckland University of Technology, Auckland, New Zealand*

Alexandros Feresidis, *Department of Electronic Electrical and Systems Engineering, School of Engineering, College of Engineering and Physical Sciences, University of Birmingham, Birmingham, United Kingdom*

Andy W. H. Khong, *School of Electrical and Electronic Engineering, Nanyang Technological University, Singapore*

B.-C. Seet, *Department of Electrical and Electronic Engineering, Auckland University of Technology, Auckland, New Zealand*

Baiying Lei, *National-Regional Key Technology Engineering Laboratory for Medical Ultrasound, Guangdong Key Laboratory for Biomedical Measurements and Ultrasound Imaging, School of Biomedical Engineering, Health Science Center, Shenzhen University, Shenzhen, China*

Dong Ni, *National-Regional Key Technology Engineering Laboratory for Medical Ultrasound, Guangdong Key Laboratory for Biomedical Measurements and Ultrasound Imaging, School of Biomedical Engineering, Health Science Center, Shenzhen University, Shenzhen, China*

Feng Jiang, *National-Regional Key Technology Engineering Laboratory for Medical Ultrasound, Guangdong Key Laboratory for Biomedical Measurements and Ultrasound Imaging, School of Biomedical Engineering, Health Science Center, Shenzhen University, Shenzhen, China*

Huang-Chang Lee, *Department of Electrical Engineering, Chang Gung University, Taoyuan City, Taiwan*

K. Veeraswamy, *QIS College of Engineering and Technology, Ongole, India*

Nguyen Q. Hanh, *School of Electrical and Electronic Engineering, Nanyang Technological University, Singapore*

Peter Gardner, *Department of Electronic Electrical and Systems Engineering, School of Engineering, College of Engineering and Physical Sciences, University of Birmingham, Birmingham, United Kingdom*

Siping Chen, *National-Regional Key Technology Engineering Laboratory for Medical Ultrasound, Guangdong Key Laboratory for Biomedical Measurements and Ultrasound Imaging, School of Biomedical Engineering, Health Science Center, Shenzhen University, Shenzhen, China*

T. Thomas, *Department of Electronic Electrical and Systems Engineering, School of Engineering, College of Engineering and Physical Sciences, University of Birmingham, Birmingham, United Kingdom*

Tianfu Wang, *National-Regional Key Technology Engineering Laboratory for Medical Ultrasound, Guangdong Key Laboratory for Biomedical Measurements and Ultrasound Imaging, School of Biomedical Engineering, Health Science Center, Shenzhen University, Shenzhen, China*

V. G. Reju, *School of Electrical and Electronic Engineering, Nanyang Technological University, Singapore*

Wanjun Li, *National-Regional Key Technology Engineering Laboratory for Medical Ultrasound, Guangdong Key Laboratory for Biomedical Measurements and Ultrasound Imaging, School of Biomedical Engineering, Health Science Center, Shenzhen University, Shenzhen, China*

X. J. Li, *Department of Electrical and Electronic Engineering, Auckland University of Technology, Auckland, New Zealand*

Xiaoli Xu, *Nanyang Technological University, Singapore, Singapore*

Yen-Ming Chen, *Institute of Communications Engineering, National Sun Yat-sen University, Kaohsiung, Taiwan*

Yeong-Luh Ueng, *Department of Electrical Engineering and the Institute of Communications Engineering, National Tsing Hua University, Hsinchu, Taiwan*

Yong L. Guan, *Nanyang Technological University, Singapore, Singapore*

Yong Zeng, *National University of Singapore, Singapore, Singapore*

Yuan Yao, *Department of Ultrasound, Affiliated Shenzhen Maternal and Child Healthcare, Hospital of Nanfang Medical University, Shenzhen, China*

List of Figures

List of Tables

List of Notations

Λ^n	Bit error at the receiver of UE n
α_n	Complex channel gain between UE n and the BS
β_n	Power allocation coefficient for UE n
$C_{SIC}(b)$	Computational complexity in terms of number of FLOPs required for decoding one bit using SIC
$C_{PIC}(b)$	Computational complexity in terms of number of FLOPs required for decoding one bit using PIC
E_b	Energy per bit
K	Processing gain
N	Total number of UEs
N_0	One-sided noise power spectral density for AWGN
N_b	Frame length in bits
N_S	Number of samples per bit
P	Total transmission power at the BS
P_j	Received power for UE j
Q	Standard Q function
S	Number of PIC stages
r_n	Received signal at the UE n
s	Superimposed signal of all the UEs transmitted by the BS
s_i	Message signal of UE i
v_n	AWGN at UE n receiver
y_n	Estimate of all other UE messages (except UE n) in the proposed receiver of UE n

List of Abbreviations

3G	3rd generation
3GPP	3rd-generation partnership project
4G	4th generation
5G	5th generation
AMC	Adaptive modulation and coding
AWGN	Additive white Gaussian noise
BER	Bit error rate
BS	Base station
CCI	Co-channel interference
CDMA	Code division multiple access
CDRT	Coordinated and direct relay transmission
CL	Closed loop
CR-NOMA	Cognitive radio inspired non-orthogonal multiple access
CSI	Channel state information
CST	Computer Simulation Technology
DL	Downlink
DPC	Dirty paper coding
DS-CDMA	Direct sequence code division multiple access
EP	Error propagation
FD	Full duplex
FEC	Forward error correction
FLOP	FLoating point Operations
F-NOMA	Fixed power non-orthogonal multiple access
FSPA	Full search power allocation
FTPA	Fractional transmit power allocation
GBPS	Giga bit per second
GSM	Global System for Mobile Communications
HARQ	Hybrid automatic repeat request
IDMA	Interleave-division multiple access
LDS CDMA	Low density spreading code division multiple access
LDS OFDM	Low density spreading orthogonal frequency division multiple access
LTE	Long-term evolution
MC CDMA	Multi-carrier code division multiple access

MCS	Modulation and coding scheme
MIMO	Multiple-input multiple-output
MMSE	Minimum mean square error
MM-Wave	Millimeter wave
MPA	Message passing algorithm
MRC	Maximum ratio combining
MUSA	Multi-user shared access
MUST	Multi-user superposition transmission
NOMA	Non-orthogonal multiple access
NOMA-BF	Non-orthogonal multiple access beamforming
OFDMA	Orthogonal frequency division multiple access
OLLA	Outer Loop Link Adaption
OMA	Orthogonal multiple access
PF	Proportional fair
PIC	Parallel interference cancelation
QoS	Quality of service
SC-FDMA	Single-carrier frequency division multiple access
SCMA	Sparse code multiple access
SI	Self-interference
SIC	Successive interference cancellation
SISO	Single-input single-output
SNR	Signal-to-noise ratio
SoC	System on chip
SU	Single user
SWIPT	Simultaneous wireless information and power transfer
TDMA	Time division multiple access
THz	Terahertz
TPA	Transmit power allocation
UE	User equipment
UMTS	Universal Mobile Telecommunications Service

1

Non-orthogonal Multiple Access: Recent Developments and Future Trends

A. Anwar, B.-C. Seet and X. J. Li

Department of Electrical and Electronic Engineering, Auckland University of Technology, Auckland, New Zealand

Abstract

The last decade has witnessed a rapid increase in the number of mobile subscribers with portable devices like smart phones and tablets to enjoy a wide range of services from simple voice to interactive multimedia. Due to scarcity of the spectrum, the current wireless systems are still unable to meet the ever-increasing subscribers' demands for bandwidth and resource-hungry applications, with a vigorous requirement of seamless connectivity, anywhere and anytime, despite the fast-growing 4th-generation systems. This motivates the wireless industry and academia researchers to define new paradigm technologies and structures for future 5th-generation (5G) wireless systems. In this regard, many enabling technologies and potential solutions are proposed, among which non-orthogonal multiple access (NOMA) is considered as a promising multiple access technology for 5G. In this chapter, we first briefly describe the classifications of NOMA. We then focus on power domain NOMA and discuss its operational principle and possible system architecture for downlink (DL). In the context of issues related to the DL NOMA system, we provide a review of recent advancements in the topic. We highlight some critical factors related to the current NOMA receiver, which inevitably limit the NOMA performance and operation. As a remedy, we suggest an alternative receiver design that may be adopted for 5G NOMA. We justify our proposal by presenting performance comparison results between the existing and proposed receiver for 5G NOMA. Finally, we provide recommendations for future research, followed by the conclusion.

1.1 Introduction

A major challenge facing telecommunication operators is the ever-increasing number of mobile subscribers and their greedy demands for high data-rate real-time multimedia services post due to the scarce spectrum. Despite the fast development of 4G mobile communication systems, they are still unable to meet the ever-increasing consumers' demands for bandwidth and resource-hungry applications, with a vigorous requirement of seamless connectivity, anywhere and anytime, and thus, the wireless communications industry and academia researchers have been compelled to define new paradigm technologies and structures to support the requirements of 5G mobile communication systems [1, 2]. Some of the staunch milestones of 5G wireless systems include $1000\times$ increase in cell throughput capacities over current 4G mobile communication systems; wide expansion in traffic and number of simultaneous connections; personalized user experience; service access from anywhere, anytime, and any device; exceptionally low end-to-end latency; and a $10\times$ increase in battery life for battery-operated services [3–5].

In order to meet the anticipated demands of 5G, many enabling technologies and potential solutions are proposed, among which ultra-densification, massive MIMO and millimeter wave have captured the attention of both academia researchers and industry [5]. Nevertheless, the role of multiple access scheme always remains a vital factor in cellular networks in order to enhance the system capacity in a cost-effective manner, while utilizing the bandwidth in such a way that overall spectral efficiency will be increased [3, 6].

The 3G mobile communication systems adopt DS-CDMA as the radio access technology, with single-user detection at the receiver side [7]. On the other hand, LTE established by 3GPP adopts OFDMA and SC-FDMA in DL and uplink (UP), respectively. The reason why LTE is opted for the OMA based on OFDMA and SC-FDMA is that it attains acceptable system throughput in scenarios of packet domain services using both time and frequency domain scheduling, while implementing a single-user detection receiver at the user terminal [8, 9]. Despite the fact that LTE presents 2–3 times capacity enhancements over current 3G systems, this gain is inadequate to meet the capacity demands of future 5G mobile communication systems [8].

Non-orthogonal multiple access with advanced transmission and reception schemes such as DPC and SIC receivers is considered as a promising

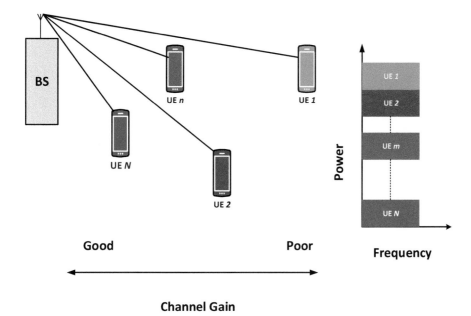

Figure 1.1 Downlink NOMA system model.

multiple access scheme to enhance system performance in both UP and DL [9, 10]. In NOMA, the transmitted signals of multiple users are multiplexed in the *power domain* at the transmitter side and demultiplexed using SIC at the receiver side [5, 6, 8, 10]. The non-orthogonality in this context means that multiple data streams for different users are superimposed on the same frequency band for transmission, but each user's signal is allocated with a different power level by the BS as shown in Figure 1.1.

In contrast to the previous generations of cellular networks that were designed and operated under the complete control of the infrastructure/ operator side, the 5G systems will aim to incorporate intelligence and enhanced processing capabilities at the UE side. In this perspective, the non-orthogonal transmission is also proposed for the cellular DL and as opposed to previous generations of cellular systems, the UEs will process the signals of multiple users at their receivers. This poses a challenge as the processing capability of UEs is still far lower than that of the BS and hence low-complexity and efficient signal-processing techniques are required at the UE side to extract its signal from a combined non-orthogonal signal of several users. Therefore, the focus of this chapter is to describe the application scenarios and related issues with DL NOMA.

1.2 Classification of NOMA Schemes

The NOMA schemes can be generally classified into two main categories, namely *code* and *power domain* multiplexing NOMA [11].

1.2.1 NOMA via Code Domain Multiplexing

The basic idea of NOMA via code domain multiplexing is similar to that of CDMA, where different users are allocated with different codes. The multi-user transmission is then carried by utilizing the same time-frequency resources. In what follows, we will briefly discuss some of the NOMA schemes via code domain multiplexing.

1.2.1.1 Low density spreading CDMA

In contrast to conventional CDMA, LDS CDMA use sparse spreading codes. The key benefit of using LDS CDMA over conventional CDMA is that it achieves less interference at the chip level due to the use of sparse spreading codes [12]. The multi-user detection at the receiver is performed with the aid of MPA. More details about LDS CDMA and MPA can be found in [13] and the references therein.

1.2.1.2 Low density spreading OFDM

Low density spreading OFDM can be viewed as an improved version of MC CDMA in which LDS codes are utilized as opposed to dense spreading codes. Similar to MC CDMA, the transmitted symbols at LDS OFDM transmitter are first mapped to LDS codes. The mapped symbols are then transmitted by using different OFDM subcarriers. A key benefit of LDS OFDM is that it allows overloading to achieve better spectral efficiency. Overloading here means that the number of transmitted symbols can be greater than the number of available subcarriers [14]. MUD at LDS OFDM receiver can be realized by using MPA [11]. Interested readers can find further details on LDS OFDM in [15].

1.2.1.3 Sparse code multiple access

Sparse code multiple access is a form of NOMA scheme which is based on a multi-dimensional codebook. It can be considered as the improved version of LDS CDMA. SCMA encoder is the most vital component which jointly performs the operations of modulation and spreading. In SCMA, every user has a predefined codebook, and as opposed to LDS CDMA, the user's data bit stream is mapped to a X-dimensional sparse codeword, where X represents the length of an SCMA codeword. Hence, the SCMA encoder spreads the user's data on X resources. The SCMA encoder is a multi-layered encoder with Z layers, where $Z = \binom{X}{Y}$, where Y denotes the non-zero dimensions of the sparse codeword. The overloading in SCMA is achieved by multiplexing Z layers to X resources. Some of the key benefits obtained by the SCMA scheme are increased throughput and connectivity, reduced detection complexity, shaping gain, better spectral efficiency, and robust link adaptation [4, 16]. Readers can find more details about SCMA in [16–18].

1.2.1.4 Multi-user shared access

The basic idea of a MUSA in the UP is that each user randomly chooses a spreading sequence from a pool of available spreading sequences. The MUSA scheme not necessarily uses the binary codewords for spreading, and in general they can be M-ary. The spread signals of the users are then transmitted using the same time-frequency resources. The receiver performs MUD by implementing codeword level SIC [19, 20].

The users are divided into different groups during DL MUSA. Within each user group, the DL MUSA scheme maps the symbols of different users to various constellations. The mapping is performed in such a way that the joint constellation of superimposed signals guarantee Gray mapping. In order to realize the joint constellation, the modulation order and transmit power among multiplexed users are utilized. The advantage of using the Gray mapping for the joint constellation is that less complex receivers (such as symbol level SIC) can be utilized to perform MUD [11, 21].

1.2.1.5 Interleave-division multiple access

Another type of NOMA is IDMA. The basic principle of IDMA is that it interleaves the chips after spreading of the symbols [22]. It is reported in [23] that under high overloading conditions with given BER constraints, IDMA achieves a SNR gain over conventional CDMA systems.

1.2.2 NOMA via Power Domain Multiplexing

The basic idea of NOMA with power domain multiplexing is to allocate different power levels to multiple users and send the superimposed signal using the same time, frequency, or code resources. MUD is performed at the receiver using SIC technique. NOMA with power domain multiplexing can be implemented in both DL and UL to increase the system capacity [24–27]. A detailed treatment of NOMA with power domain multiplexing will be given in the next section.

In the rest of the chapter, we focus on the NOMA via power domain multiplexing, and for the sake of simplicity, by NOMA, we always mean power domain NOMA.

1.3 NOMA

In this section, we will discuss the basic principle, transceiver architecture, and some motivations to adopt NOMA as a MA scheme for 5G.

1.3.1 Basic Principle

Let us consider a scenario of DL cellular transmissions in which a BS is simultaneously transmitting signal to all the UEs in a cell, as shown in Figure 1.1. Assume that there are a maximum of N UEs in a cell. Let α_n be the complex channel gain between UE n and the BS. Further assume that the total transmission power at the BS is constrained to P, so $\sum_{n=1}^{N} \beta_n P = P$, where β_n is the power allocation coefficient for UE n. Suppose $|\alpha_1|^2 \geq |\alpha_2|^2 \geq \cdots \geq |\alpha_N|^2$ and hence the power allocation coefficients can be sorted as $\beta_1 \leq \beta_2 \leq \cdots \leq \beta_N$. The superimposed signal of all the UEs transmitted by the BS is then given by [10, 28]:

$$s = \sum_{i=1}^{N} \sqrt{\beta_i P} s_i \qquad (1.1)$$

where s_i is the message signal for UE i. The received signal at the nth UE can be represented as [28]:

$$r_n = \alpha_n s + \nu_n = \alpha_n \sum_{i=1}^{N} \sqrt{\beta_i P} s_i + \nu_n \qquad (1.2)$$

where ν_n is the AWGN.

The optimal decoding order for the users is in the order of increasing channel gains. Hence, in this case of channel gains order, the UE N having the least channel gain will decode its signal in a straightforward manner, as it will treat all the signals from other UEs as noise. At the receiver of UE n, $1 \leq n \leq N$, all the message signals for UEs $n+1$, $n+2, \ldots, N$ will be considered as interference and they will be removed in a successive fashion using SIC. After the removal of the interference from higher order UEs, it will decode its own message signal by treating messages for lower order UEs 1, 2, ..., $n - 1$ as noise [10].

1.3.2 NOMA Transmitter and Receiver Architectures

The transmitter and receiver architectures for DL NOMA system are shown in Figure 1.2. At the transmitter side, the input data bits are converted into the codewords by the FEC encoding block. The output of the FEC block is interleaved by the interleaving block, which is then fed into the modulation block to obtain the transmitted signal s_i for UE i, where $1 \leq i \leq N$.

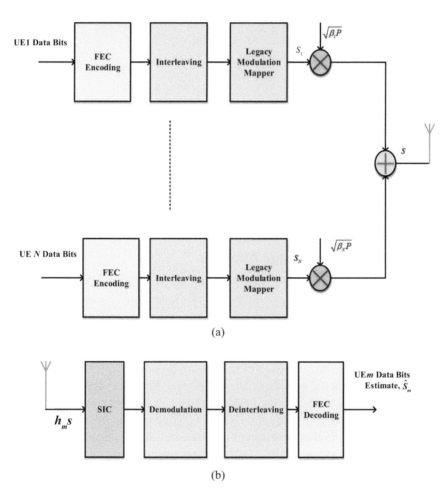

(a)

(b)

Figure 1.2   (a) NOMA transmitter and (b) UE receiver.

In order to realize a power domain multiplexing, the message signal of every UE is multiplied by the allocated power, i.e., s_i is multiplied by $\sqrt{\beta_i P}$. The transmitted NOMA signal s is then constructed by adding the message signals of all the UEs.

At the receiver side, SIC is performed to remove the intra-user inter-ference followed by demodulation. The output of the demodulation is then fed into the deinterleaving block. Finally, the FEC decoding is performed to obtain an estimate of the message signal of UE.

1.3.3 Motivations to Adopt NOMA as MA Scheme for 5G

The following are some of the key factors that make NOMA a promising candidate over OMA to adopt as a MA scheme for 5G:

- Enhanced cell throughput
- Increased capacity
- Efficient spectrum utilization
- Superior outage performance
- Better energy efficiency.

1.4 Review of Some Recent Developments for NOMA

The review presented in this section is performed by considering the follow-ing issues related to DL NOMA.

1.4.1 Throughput and Outage Analysis

In the literature, the DL NOMA is investigated with SIC receiver at the UE side. In [6, 29, 30], the performance of the NOMA technique is evaluated with SIC receiver by performing system-level simulations. In [29], the authors suggested NOMA as a promising multiple-access scheme for future cellular systems. The utilization of additional power domain by NOMA for user multiplexing, the expected increase in the device processing capabilities for future systems, and robust performance gain are the key motivations behind this proposal. The authors performed system-level simulations and compared the performance between NOMA and OFDMA. They presented the compar-ison in terms of the number of users and the overall cell throughput under frequency-selective/wideband scheduling and power allocation. The authors reported a performance gain of 30–35% in the overall cell throughput for NOMA over OFDMA.

In order to achieve a further performance gain using NOMA, the authors proposed combination of NOMA with MIMO and multi-site extensions as possible future extensions. The authors adopted the SIC receiver for NOMA as a standard decoding scheme at the UE side. This choice implies that the performance of NOMA is directly dependent upon the decoding performed by the SIC receiver. Hence, a decoding failure by SIC results in performance limitation of NOMA. However, in the comparisons presented in [29], they assumed a perfect decoding for SIC and hence, neglected the impact of decoding failure and EP of SIC on the performance of NOMA.

In [31], the authors considered a single-cell case for DL NOMA and modeled the cell using a circular disc geometry of radius R_D. They assumed that the users are distributed randomly with uniform distribution in the circular disc. In order to evaluate the performance of DL NOMA, they presented two scenarios for user's rate and calculated the corresponding performance metric. In the first scenario, the QoS requirements of the user are utilized to determine the data rate. They considered two scenarios: the idealistic suppression of the interference caused by all the higher order users and the ability of the NOMA to satisfy the QoS requirements of the user. In the second scenario, they calculated the user's rate opportunistically based on his channel conditions. The outage probability is considered as a performance metric in the first scenario, while for the second scenario, the achievable ergodic sum rate by NOMA transmission protocol is adopted. The analytical and numerical results show that the NOMA is able to achieve better outage and ergodic sum rate performance than OMA. However, there are two critical factors that can adversely affect the outage performance of the system, namely the selection of user's targeted rate and the choice of power allocation coefficient. The reason is that a wrong or inappropriate choice of a power allocation coefficient will result in a decoding failure for SIC and consequently this will lead to high outage probability.

In [6] and [30], to incorporate the practical prospects of the cellular systems into their system-level simulation evaluations of DL NOMA with SIC receiver, the authors combined LTE radio interface functionalities such as time/frequency domain scheduling, AMC, HARQ and Outer Loop Link Adaption (OLLA) with NOMA specific operations such as dynamic power allocation. A proportional fairness scheduler was used to achieve the multi-user scheduling. Furthermore, dynamic TPA scheme was used to distinguish the users in power domain. The HARQ mechanism was used if decoding error occurs.

1.4.2 Power Allocation and User Grouping

In [6], the authors also presented an algorithm to implement the complete NOMA signal generation in the DL. The authors used FSPA and FTPA to distribute the power among NOMA users. The signaling cost related to SIC decoding order and power assignment ratios will be enhanced with the use of dynamic TPA. FSPA scheme for NOMA achieved the best performance but it is computationally complex, as all combinations of power allocation are considered.

In contrast, FTPA provides a low-complexity method for power allocation among NOMA users. The power decay factor (whose value ranges from 0 to 1) in FTPA is responsible for power allocation relative to the channel gain and is a parameter that requires optimization in order to meet the required performance targets. In [6], the authors choose the value of power decay factor as 0.4. The impact of varying the power decay factor during FTPA is not considered by the authors, and the chosen power decay factor may not be a suitable choice for NOMA under some other communication scenarios such as users having comparable channel quality, equal power allocation, or having fairness constraints.

The consequence of user grouping for NOMA with the power allocation is also studied in [30]. The purpose of user grouping is to group those users who have a significant difference in their channel gains, so that the amounts of power allocated to them using TPA or FTPA will differ substantially. The authors reported a performance gain of 35–40% in [6, 30] over OFDMA in terms of cell throughput and user rate under power allocations schemes of TPA and FTPA with multi-user scheduling. However, in both the works of [6] and [30], the authors assumed a significant difference in channel conditions of the scheduled users, which may not always be true in practice. There exists a finite probability that at some of the transmission times, the scheduled users may not differ significantly in terms of channel gains, which implies negligible differences among their power allocations. This will lead to difficulty in grouping/pairing the users. Further, the SIC receiver is power sensitive [32, 33], and thus the performance of SIC will degrade intensely if two or more users have similar or comparable power. Consequently, by making an assumption of holding a significant difference in channel gains of the scheduled users, the authors of [6] and [30] ensure different power allocation to multiple users in order for SIC to work properly and hence neglected the impact of SIC decoding failure that may happen when there is no significant difference in channel conditions of the scheduled users.

The impact of user pairing on the performance of NOMA was also investigated in [33]. The authors considered two DL systems namely F-NOMA and CR-NOMA, and then demonstrated the effect of user grouping on their performance. In the F-NOMA system, the authors considered a case of two users and studied the impact of user grouping on the system performance. The authors analyzed the system from the perspective of achievable sum rate (probability that F-NOMA achieves a lower sum rate than conventional OMA) and user's individual rate (probability that F-NOMA achieves a higher rate than OMA) and presented their closed-form analytical expressions. Their results depicted a superior performance of F-NOMA over OMA. However, they reported that the careful user grouping is required to obtain this performance, and in order to attain this performance benefit, the users with significant difference in channel qualities should be paired together. This implies that the performance gap achieved in F-NOMA over OMA is directly dependent upon the fact that how large the difference in channel qualities for the grouped users is. If the grouped users have similar channel conditions, then the benefit of using NOMA will be rather limited. This performance limitation is due to the inherent constraint attached to the use of SIC receiver with NOMA, i.e., how efficiently the SIC receiver performs decoding and detection of multi-users hugely relies the spread of user powers [34, 35]. This was the reason why authors emphasized maintaining a significant difference between the channel conditions of the grouped users in order to achieve the performance gap over OMA by using the F-NOMA scheme.

In the CR-NOMA scheme proposed by the authors in [33], the user having lower channel gain is considered as the primary user, while the user with strong channel is regarded as a secondary user. The secondary user has been allocated power opportunistically and is permitted to transmit on the channel of the primary user only on the condition that it will not cause the performance of the primary user to degrade adversely. This condition is fair enough to maintain power difference between the primary and secondary users and is sufficient for SIC technique employed at the primary user's receiver to decode his data from DL signal received in the CR-NOMA scheme. Therefore, the core purpose of the CR-NOMA scheme is to serve the primary user (user having poor channel gain) by allowing the secondary user to transmit on the condition that its transmission will not harm the primary user's signal. Consequently, this arrangement will achieve a higher sum rate than OMA. Apart from the requirement of having significant difference in channel gains to group the users and allowing the secondary user to transmit on the primary user channel under the condition mentioned above so that the SIC will work properly, in both of the proposed schemes, the authors investigated the user grouping impact by considering the case of two users. This makes grouping simple as only two users are picked from a large set of users. However, it is not a trivial problem to group more than two users (e.g., five) by maintaining the significant difference in channel conditions.

1.4.3 Fairness in NOMA

The issue of fairness achieved by the DL NOMA protocol is addressed in [5]. They considered a problem of distributing a total available power at the BS among the NOMA users and investigated its impact on the fairness under two situations: (1) determination of the user's data rate under perfect knowledge of CSI and (2) the user's date rate is determined on the average CSI, i.e., a fixed targeted data rate. In first case, they adopted max-min fairness that aimed to maximize the minimum achievable rate by the user as a performance comparison metric, while in the second case, the outage probability was considered as a performance metric. In both cases, the authors considered the user's rate and outage probability as a function of power allocation coefficient and formulated them as a non-convex optimization problem. The authors proposed low-complexity iterative power allocation coefficients as optimal solutions in both cases. This work can be regarded as an extension of [31] by finding an optimal power allocation coefficient under two assumptions of perfect and average CSI for the users, with the focus on investigating its impact on the fairness achieved by DL NOMA with SIC. The numerical results showed that with the use of the proposed optimal power allocation schemes, NOMA maintains high fairness and can achieve superior performance in terms of rate and outage probability as compared to TDMA and fixed power allocation in [31]. However, this fairness will only be maintained when the power is allocated appropriately and deviation from this condition will result in high outage probability due to the decoding failure by the SIC receiver.

1.4.4 MIMO NOMA

In [36], the authors considered a DL MIMO cellular system considering NOMA with SIC as an underlying multiple access scheme. The authors proposed an open-loop type random beamforming scheme which results in the reduction of required feedback information from UEs. In the proposed beamforming scheme, the multiple beams are generated at the BS. In each of the generated beams, multiple users are superimposed using the NOMA transmission protocol. This implies that if there are a total of four scheduled users and two beams are generated at the BS, then, two UEs are sent in one beam while the remaining two are sent in the other beam using NOMA transmission. The authors named the transmission of multiple users using NOMA in one beam intra-beam superposition coding.

The implementation of this proposed random beamforming scheme with the NOMA protocol introduces two types of interferences to be canceled at the UE receiver. The first is the inter-beam interference and the second is intra-beam interference. In order to minimize the impact of inter-beam interference, the authors used MMSE as a spatial filtering technique. After suppressing the inter-beam interference, the SIC receiver is used to decode the intra-beam interference, which is due to the transmission of multiple users in a particular beam. The authors performed simulation of the proposed system and presented their analysis in terms of cell-edge and sum user throughputs, and reported a superior performance of the proposed system over OMA.

However, perfect channel knowledge was assumed in their simulations. Moreover, a maximum of two beams were considered for transmission and therefore, they neglected the impact of inter-beam interference with an increasing number of beams on the proposed system performance. Further, the author provided no information on how many users were multiplexed in a single beam using the NOMA protocol. This is a critical consideration to evaluate the performance of SIC receiver which was suggested by the authors for the mitigation of intra-beam interference.

The system proposed in [36] is further investigated in [9]. In contrast to [36], they considered imperfect channel and the influence of channel estimation errors on the overall system throughput by performing system-level simulations. In order to minimize the impact of channel estimation errors on the performance of the system, they presented a rate transmission back-off algorithm. In this algorithm, they introduced a parameter α which was responsible for controlling the back-off amount. The rate and α were related inversely to each other, and as a consequence, by tuning α, the rate is adjusted accordingly. This technique implies that in the presence of channel estimation errors, simply lower down the transmission rate by increasing the value of α. The apparent drawback of this scheme is that by increasing α to a larger value, the transmission will be occurring at a very lower rate as compared to the channel capacity and hence it is not a bandwidth efficient scheme. Further, the authors assumed perfect decoding for the SIC receiver, i.e., the intra-beam interference is perfectly removed. This is a critical assumption because potential decoding error propagates in the SIC scheme and it will increase the intra-beam interference and abruptly influence the performance of the proposed NOMA system with random beamforming.

In [8], the authors considered DL NOMA and investigated the impact of CL single-user (SU) MIMO on its performance. In order to make their analysis more practical, they included two performance-limiting factors in their system, namely the EP for the SIC receiver and UE velocity. They proposed two EP models for the SIC receiver. While proposing the EP models for SIC and also in all their performance comparisons, the authors considered only two NOMA users, with UE1 having a lower power level as compared to UE2. Therefore, UE2 always decodes its signal directly by considering UE1 signal as noise, while at UE1, the signal of UE2 is always decoded first and then this decoded signal will be used by UE1 to decode its own signal. Hence, in this scenario, the impact of EP is relevant only to the UE1 receiver, because there is merely one higher order user exists. In order to model the impact of EP on UE1 decoding, the authors proposed worst-case and realistic EP models. In the worst-case model, they assumed that the incorrect decoding of UE2 at stage 1 of the UE1 receiver always results in an unsuccessful decoding of UE1, while in the realistic EP model, they treated the decoded signal of UE2 from the output of stage 1 of the UE1 receiver as an interference signal and used it to decode the signal of UE1 at stage 2 of its receiver.

The authors of [8] presented their results in terms of attaining gains in average cell and cell edge throughputs as compared to OFDMA, the degradation in throughputs (average cell and cell edge) considering the proposed EP models, and the performance gain in throughput relative to UE velocity. The authors reported that using the EP models, the degradation in NOMA gain is nearly 2.32 and 1.96% for cell average and cell edge user throughputs, respectively. However, there are some factors that can pose challenge in the generalization of these results. First, it is anticipated that by increasing the number of NOMA users beyond two, the NOMA gain will be reduced by a much greater percentage as reported by the authors. The reason is that the SIC receiver decodes in successive manner [34], and if a decoding failure occurs for any of the higher order decoding user(s), then that error will be propagated through the lower order user(s) until to the desired UE. Thus, even with the proposed EP models, the error will still tend to propagate to the lower order decoding users and its impact will surely increase if the number of NOMA users increases, resulting in a higher probability of unsuccessful decoding. Further, they did not provide any mathematical analysis for the performance parameters used that relate and justify their system-level simulation results for all possible scenarios. Therefore, without an analytical study, it is not possible to generalize their results beyond two UEs.

The authors in [37] applied MIMO to DL F-NOMA. There are M transmit antennas at the BS, while each user is equipped with N receive antennas. The total users are divided into M clusters and each cluster has K users. The authors proposed a novel structure of precoding and detection matrices. The precoding and detection matrices were used to suppress the inter-cluster interference, while the intra-cluster interference is minimized with the aid of the SIC technique, assuming the perfect decoding of all the higher order user. The authors assumed that the BS has no knowledge about the global CSI, i.e., it does not know the channel matrices of all users, which minimizes the system overhead at the BS due to CSI acquisition. This implies that the BS is transmitting data to users without manipulation and hence they choose the precoding matrix as an identity matrix. Following this choice of precoding matrix, the authors derived the detection matrix and presented maximum ratio combining (MRC) as one possible implementation of it. The authors proved that both MIMO-OMA and MIMO-NOMA are capable of achieving the same diversity gain but the MIMO-NOMA scheme is more spectrally efficient and hence results in a larger sum rate. In order to reduce the system complexity, the authors proposed user pairing as a promising solution and investigated its impact on the performance. They reported that with the aid of user grouping, MIMO-NOMA will be able to attain better outage probability performance as compared to conventional MIMO-OMA. This article presents concrete analysis and elegant mathematical expressions for outage probability and sum rate gap between MIMO-NOMA and MIMO-OMA. The authors grouped two users which makes its implementation easier. However, if grouping a larger number of users is desired, two major problems will arise: first, the process of user pairing will become complex, and second, the problem of power allocation to meet the conditions for NOMA to remain operative will be challenging.

1.4.5 Massive MIMO NOMA

In order to further increase the spectral efficiency of NOMA for 5G, the authors in [19] proposed to combine NOMA and massive MIMO technologies. In this short paper, the authors presented system analysis of massive MIMO with NOMA in terms of capacity and they give the ergodic sum rate achieved by the massive MIMO system with NOMA. In addition, the energy efficiency analysis of massive MIMO with NOMA was performed. However, the authors did not present any performance comparison results to justify their analysis. Further, they did not provide system modeling of massive MIMO with NOMA, which is essential to perform any analysis.

A more sophisticated and comprehensive design of NOMA system with massive MIMO is presented in [39]. In their proposed scheme, the authors analyzed the massive MIMO NOMA system by decomposing it into different separated SISO NOMA channels. The authors presented the performance analysis of their proposed scheme under two scenarios—with perfect user ordering and limited feedback. The authors presented their performance analysis results in terms of outage sum rates and outage probabilities.

1.4.6 Cooperation in NOMA

In the work of [40], the authors proposed a cooperative DL NOMA system having a broadcast channel with one BS. This cooperation is among the NOMA users that aim to improve the decoding of individual user's data by suppressing the multi-user interference. The main motivation to introduce the cooperative NOMA is to take advantage of the inherent characteristic of the NOMA protocol in which the users having strong channels are aware in advance of the messages of the other users. The proposed system is divided into direct transmission and cooperative phases. The direct transmission phase works exactly in the same manner as the conventional DL NOMA operates. In the cooperative phase, the users cooperate with each other by broadcasting the decoded messages of the higher order users with the aid of time slots. At each time slot, one user broadcasts the combination of messages equal to the number of higher order users from him. This process starts from the user having the strongest channel till before the user having the worst channel quality. All the users utilize the cooperative information sent by the users having channel qualities greater than theirs to first decode the messages of the higher order users and then at the second step they decode their own messages. This implies that in the first time slot, the user having the strongest channel will transmit cooperative information, while in the last time slot, the user having the second worst channel will transmit the cooperation message.

The implementation of this cooperative NOMA scheme has immensely enhanced the reception reliability. However, allowing all the users to cooperate will increase the system complexity and overhead to develop coordination among multiple users, which poses a challenge for the practical implementation of this scheme. As a remedy, the authors suggested user grouping as one possible solution, in which users situated in a cell are divided into different groups, and the cooperation is implemented between these different groups. The analytical and numerical results reported in terms of outage probability and average sum rate difference by the authors demonstrate the better performance over non-cooperative NOMA and OMA in all SNR regimes. The cooperative scheme proposed by the authors for NOMA essentially helps the users to improve the decoding of their own messages by eliminating or minimizing the multi-user interference from the other users. Therefore, one perspective to view this proposed scheme is that it aims to improve the estimation of individual user's data by removing the multi-user interference with the aid of cooperation. Hence, it motivates one to estimate the multi-user interference at the receiver of each user with relatively lower algorithmic and implementation complexity, so that it will be targeted at improving the detection performed by the SIC.

1.4.7 NOMA for Relaying Networks

In [41], the authors propose a design of a NOMA system for multiple-antenna relaying networks. In their proposed scheme, the authors considered a DL cooperative cellular scenario, where they assumed that the BS is communicating with the mobile users with the aid of relays only, i.e., there is no direct link between the BS and mobile users. They evaluated the performance of their proposed scheme in terms of outage probability and provided its closed-form expression under the considered network setting. The authors reported that the NOMA can achieve better outage performance than conventional OMA under the condition when the relay is located closer to BS. However, when the relay is located closer to mobile users, the OMA is able to achieve better outage performance. Hence, the placement of relay nodes could be a vital factor to obtain the performance gain of NOMA over OMA and needs further investigation for the considered system model in [41].

In the work of [42], the authors considered NOMA for CDRT. In their proposed scheme, the BS is communicating directly with UE1 and relay, whereas UE2 receives data from the relay, as there is no direct link between the BS and UE2. The authors provided the performance comparisons in terms of outage probability and ergodic sum capacity and derived their closed-form analytical expressions. They show that under the considered network scenario, their proposed NOMA in CDRT achieves a significant gain over NOMA in non-CDRT NOMA.

1.4.8 NOMA and Simultaneous Wireless Information and Power Transfer

The authors in [43] applied simultaneous wireless information and power transfer (SWIPT) to NOMA networks with randomly distributed users. In particular, they proposed a cooperative SWIPT NOMA, where NOMA users closer to the BS act as energy-harvesting relays for far NOMA users and help to improve their reliability in a cooperative fashion. They proposed three strategies to select users based on their distances from the BS, and provided closed-form expressions for outage probability and system throughput for the three proposed user selection schemes. The main benefit of the proposed cooperative SWIPT scheme is enhanced reliability for far users and overall improved spectral and energy efficiency.

A summary of the literature review performed above is given in Table 1.1.

Table 1.1   Summary of literature review

Issues Investigated	Highlights	References
Throughput and outage analysis	Performed system-level simulations to present performance comparison between NOMA and OMA choosing cell throughput and user rate as performance parameters.	[6, 29, 30]
	The authors provided performance analysis of NOMA with randomly deployed users. The analytical expressions for user's rate under two scenarios are derived and they presented analytical and simulation results for user's rate to show the superiority of NOMA over OMA.	[31]
	The authors proposed combining LTE interface functionalities with NOMA-specific operations in order to incorporate the practical aspects of the cellular systems.	[6, 30]
Power allocation and user grouping	The authors investigated the impact of user pairing in NOMA by using fixed and cognitive radio inspired power allocation schemes. The achievable sum rate and user's individual rate were considered as performance metrics and the authors provided analytical and simulation comparisons between two considered power allocation schemes for NOMA and OMA.	[33]

	The authors provided an optimal power allocation for NOMA in order to achieve fairness. The performance comparison is presented in terms of rate and outage probability between proposed power allocation scheme for NOMA, TDMA and F-NOMA.	[5]
	FSPA, FTPA and TPA are used by the authors to group users in DL NOMA. With the aid of user grouping enabled by the power allocation in NOMA, the authors reported a performance gain over OMA.	[6, 30]
Interference handling in MIMO	The authors proposed a random beamforming scheme for transmission in DL NOMA. The inter-beam and intra-beam interferences are introduced due to the implementation of this scheme. Under the assumption of perfect CSI, the authors used the MMSE technique to suppress inter-beam interference, while SIC is used to minimize the intra-beam interference.	[36]
	The authors extended the work of [15] under imperfect channel and considered the impact of channel estimation errors on overall system throughput. In order to minimize the influence of channel estimation errors, they introduced a transmission back-off algorithm.	[9]
	The authors considered DL NOMA and included the effect of EP for SIC and UE velocity in their analysis. They proposed two EP models for SIC. The authors presented the performance comparison between NOMA and OMA considering the proposed EP models and UE velocity impact.	[8]
	The authors applied MIMO to DL F-NOMA. They proposed a novel design for precoding and detection matrices to suppress the inter-cluster interference whereas the intra-cluster interference is minimized with the aid of the SIC technique.	[37]
Combining massive MIMO with NOMA	The authors proposed to combine massive MIMO with NOMA to increase the spectral efficiency of NOMA.	[38]
	The authors proposed a design of massive MIMO for NOMA.	[39]
Cooperation	A cooperative DL NOMA system having a broadcast channel with one BS is proposed. This cooperation is among the NOMA users which aims	

(Continued)

Table 1.1   Continued

	to improve the decoding of individual user's data by suppressing the multi-user interference. The analytical and simulation results are provided in terms of outage probability and average sum rate difference by performance comparison metrics.	[40]
Relaying	The authors considered a cellular network with NOMA transmission protocol and relay, such that there is no direct link between the BS and mobile users. They considered outage probability as a performance metric and provided its closed form-expression.	[41]
	The authors proposed a CDRT scheme in which near BS is directly communicating with the near UE, whereas the far UE is receiving data from BS with the aid of a relay. They presented the performance comparisons in terms of outage probability and ergodic sum capacity.	[42]
Wireless information and power transfer	The authors proposed a cooperative SWIPT scheme to improve the reliability of far UEs where near UEs act as energy-harvesting relays for them. They evaluated the system performance in terms of outage probability and throughput under the proposed schemes.	[43]

1.5 Performance-limiting Factors for Existing NOMA

As mentioned earlier, in the literature, an SIC receiver is used at the UE receiver for DL NOMA transmissions. However, there are several potential issues of SIC that can adversely affect the NOMA performance in DL and restrict NOMA functionality. The major shortcomings of the SIC receiver include:

- Successive interference cancellation is power sensitive meaning that if two or more NOMA users have same/comparable power, then the performance of SIC will degrade dramatically and this results in the reduction of overall system performance [34].
- The performance of the SIC receiver is highly dependent upon the correct decoding of the first user. In case the first UE is not decoded correctly, this error in decoding will propagate successively to lower order UEs and hence deteriorate the system performance [6, 34].
- The SIC receiver removes the multi-user interference in a successive fashion. Consequently, increasing the number of scheduled users will increase the decoding delay faced by the lower order users [6].

- The use of SIC receiver at the UE side requires appropriate power and rate allocation to scheduled NOMA users and failing to comply with these requirements results in high outage probability [31].

In order to alleviate the aforementioned performance-limiting factors for NOMA, we propose an alternate receiver structure in the next sub-section.

1.5.1 Proposed PIC-based Receiver Structure

This section describes the structure and operation of the proposed PIC-based receiver at UE for 5G DL NOMA transmissions. The block diagram of the proposed PIC-based receiver for UE n is shown in Figure 1.3.

The received composite signal at the nth UE is given by (2). The proposed PIC-based receiver will perform decoding in two steps. In the first step, the proposed receiver will remove the multi-user interference caused by the transmission from users 1, 2, $\ldots, n-1, n+1, \ldots, N$. In contrast to SIC, the proposed receiver will mitigate the multi-user interference in a parallel fashion. After removing the interference from other users, at the second step, the receiver will decode the message of UE n.

In order to illustrate the multi-user interference cancellation procedure, let us consider an example of UE n with received signal r_n. Let $\hat{s}_i$ be the estimate of message signal of UE i after decoding. Then, if we sum the estimates of all the i UEs and subtract it from the original received signal r_n, we will obtain an estimate of the message signal for UE n. Hence, the signal at the input of the decoder for UE n can be written as:

$$y_n = r_n - \sum_{i=1, i \neq n}^{N} \hat{s}_i, i = 1, \ldots, n-1, n+1, \ldots, N \qquad (1.3)$$

The decoder of UE n will perform decoding on y_n to get the estimate of the message signal for UE n.

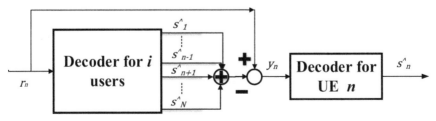

Figure 1.3 Proposed receiver structure for UE n.

Filter bank design for decoder of i users

Filter bank design is one of the possible approaches for implementing a block for decoder of i UEs to cancel the multi-user interference in a parallel manner. The composite signal r_n is fed into the matched filter input of each UE i, where $i = 1, 2, \ldots, n - 1, n + 1, \ldots, N$. The output from each matched filter is then used to get the estimate of the transmitted message signal for that UE. The resulting probability of bit error at the receiver of UE n (Λ^n) in an AWGN channel is given as [32, 34]:

$$\Lambda^n = Q\left(\left[\frac{N_0}{2E_b}\left(\frac{1 - (N - 1/K)^{S+1}}{1 - (N - 1/K)}\right) + \right.\right.$$

$$\left.\left. + \frac{1}{K^{S+1}}\left(\frac{(N - 1)^{S+1} - (-1)^{S+1}}{N}\frac{\sum_{j=1}^{K} P_j}{P_n} + (-1)^{S+1}\right)\right]^{-1/2}\right) \tag{1.4}$$

where P_j is the received power for UE j, N is the total number of UEs, K is a processing gain, S is the number of PIC stages, and $Q(.)$ is a standard Q function.

The proposed PIC-based receiver solves the problems of decoding failure in SIC due to equal or comparable power allocation for two or more NOMA DL UEs, successive EP, decoding delay, and dependency on the correct decoding of the first UE. All of these characteristics of the proposed receiver make it a stronger choice to adopt for the receiver side of UE in 5G DL NOMA transmissions.

It should be noted that the proposed receiver uses the same approach to cancel interference in parallel as used in PIC receivers designed for CDMA systems. The CDMA systems assign unique codeword to every user [44]. The codewords are utilized at the receiver side to calculate cross-correlation between the received signal and the codeword of each user in parallel by using matched filter bank. In contrast, NOMA uses power domain to multiplex different users in the same resource (time, frequency, or code), and hence the NOMA users may not have unique signature signals as are present in CDMA systems. This poses a challenge to design a decoder for i users in Figure 1.3, particularly when time/frequency resources are utilized for NOMA transmissions. Further, the proposed receiver is designed for AWGN channel and designing it to compensate for fading and path loss requires further study and analysis.

1.5.2 Performance Comparison

In this section, the analytical performance comparisons between PIC and SIC are presented. Apart from the fact that PIC solves the problems posed by SIC, the analysis also serves to promote the proposed PIC-based receiver structure as an alternative to SIC for 5G DL NOMA based on its better performance.

The first comparison is presented between probability of bit error and number of UEs, as shown in Figure 1.4. The results are obtained with 30 UEs for both SIC and PIC. Each set of results is collected by keeping constant SNR of 10, 15, and 20 dB. The results show that by increasing the number of UEs beyond 5, there is negligible impact of increasing SNR for SIC receiver; while for PIC, the performance gain is appreciable up to 15 users for all three considered values of SNR.

The second set of results is presented in terms of probability of bit error and SNR, as shown in Figure 1.5. This figure shows the analytical BER performance of SIC and PIC along with simulation results for NOMA with SIC and PIC receivers for the case of three users. The power allocation coefficients β_i, where $i = 1, 2, 3$ for three NOMA UEs are considered to be random variables (due to the random nature of the channel), uniformly distributed between (0,1). In this way, we considered and incorporated the case of equal or comparable power allocation for two or more NOMA users in our simulations, as there exists a finite probability that two or all three NOMA users can be allocated with the same power coefficient, which implies

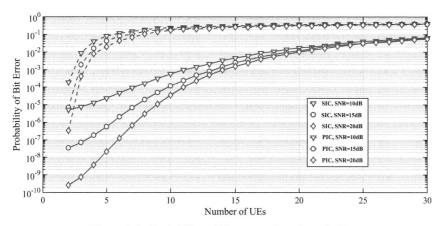

Figure 1.4 Probability of bit error and number of UEs.

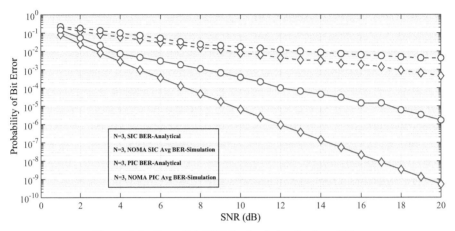

Figure 1.5 Downlink NOMA simulation for three UEs.

that they have the same or comparable channel quality. This choice for considering the power allocation coefficient as random variable makes it more realistic and general. The analytical results for SIC and PIC receivers provide a performance bound for NOMA simulation. The simulation results show that NOMA with PIC receiver at UE performs better than NOMA with SIC in all SNR regions.

It can also be observed form Figure 1.5 that there is a larger gap between the analytical and simulated performances for PIC, whereas for SIC the analytical and simulated performances are closer. The reason behind this gap between analytical and simulated performance for PIC is that the analytical results were obtained by considering (1.4), where it is assumed that the estimate of each user is perfect and unbiased, which results in perfect removal of multi-user interference. Equation (1.4) acts as the analytical performance bound for PIC.

However, while performing the simulation for three users, the estimates obtained for the users' data in parallel at the receiver of some UE are not perfect, which results in residual interference to exist. The existence of this residual interference results in the larger deviation from the analytical performance. On the other hand, it can be observed that there is no large difference between the simulated and analytical performance of SIC. The reason for this is that SIC technique does not remove the multi-user interference in one step; rather the interference is removed successively.

In contrast to PIC, at each step of decoding in SIC, only one user is decoded and removed, while the interference from all the other remaining UEs remains. This implies that there always exists an interference component in the signal at each step of decoding till last UE. Therefore, the analytical and simulated results do not differ largely for SIC.

The final set of comparison between the SIC and PIC techniques is presented in terms of computational complexity and is shown in Figure 1.6. The computational complexity can be defined as the number of FLOPs required to decode one bit. A FLOP can be an operation implementing a multiplication or addition, while more complex operations can be regarded as multiple operations. For analysis purposes, we use the computational complexity per bit of decision derived for the SIC and PIC receivers in [34], given by:

$$C_{PIC}(b) = NL\left[S\left(6KN_s + 7\right) - 4KN_s - 1\right] \tag{1.5}$$

$$C_{SIC}(b) = L\left[N\left(8KN_s + 12\right) - 6KN_s - 7\right] + \frac{N}{N_b}\left[2N_b + \log_2\left(N\right) + 1\right] \tag{1.6}$$

where $C_{PIC}(b)$ and $C_{SIC}(b)$ are computational complexities in terms of number of FLOPs required for decoding one bit using the PIC and SIC receivers, respectively, L is the maximum number of multipath signals processed by correlators at the receiver, K is a processing gain, N is the total

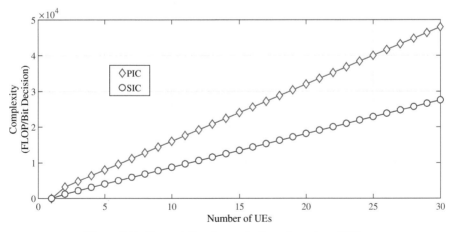

Figure 1.6 Computational complexity and number of UEs.

Table 1.2   Simulation parameters

Parameter	Value
Number of bits, N_b	100
Number of samples per bit, N_S	10
Number of multipath signals, L	10
Processing gain, K	1
Number of UEs, N	30
Number of PIC stages, S	2

Table 1.3   Computational time of PIC for different mobile SoCs [45–47]

SoC	MFLOP/sec	T (μs)
Samsung Exynos 3110	17.7	2800
Nvidia Tegra 2	54.4	900
Texas Instruments OMAP 4460	75	653
Qualcomm Snapdragon 600	540	90.7
Samsung Exynos 5 Octa 5410	627	78
Qualcomm Snapdragon 800	1034	47.3

number of UEs, N_S is the number of samples per bit, and N_b is the frame length in bits.

The computational complexity is computed for the parameters stated in Table 1.2. The increased computational complexity for PIC, as depicted in Figure 1.4, is due to the parallel structure of this receiver. The performance gains in Figures 1.4 and 1.5 were achieved at the cost of higher computational complexity.

The computational time T (in μs) for PIC is calculated for these SoCs (Table 1.3). The maximum considered computational complexity for PIC with $N = 30$ is approximately 4.9×10^4 FLOP, as shown in Figure 1.6. The different values of T show that the current and the upcoming generations of the smartphones will have a capability to implement PIC-based receiver for 5G DL NOMA.

1.6 Future Research Direction

1.6.1 Modulation and Coding Scheme

Modulation and coding are the essential components of the transmitter in communication system. In order to realize the gains of NOMA at the physical layer, a practical design for MCSs are required [24], which aim to achieve further improvement in spectral efficiency and BER. Since the users are multiplexed in power domain and there is no orthogonality among the multiplexed users (sharing same time, frequency or code resources) that can be exploited to minimize the intra-user interference in NOMA, the design of MCS is challenging for NOMA and is a research challenge to address.

1.6.2 Hybrid MA

Non-orthogonal multiple access transmission protocol is CCI limited. Therefore, it will be beneficial to use hybrid MA by combining NOMA and conventional OMA schemes [43]. It is anticipated that the future cellular networks will implement hybrid MA schemes. MUST is a type of hybrid MA scheme utilizing both OFDMA and NOMA. In MUST, NOMA is applied whenever there is a significant difference in the channel gains of the users. Hence, it would require further investigations to propose new hybrid MA schemes for future cellular systems by combing NOMA with conventional OMA and newly proposed MA schemes for 5G networks [24].

1.6.3 Imperfect CSI

The assumption of perfect CSI is common in most of the literature related to NOMA. However, in reality, it is hard to obtain a perfect CSI due to the channel estimation errors. The attempt to enhance the channel estimation with the aid of more pilot signals is spectrally inefficient. Therefore, evaluation of NOMA system reliability under the condition of imperfect CSI is an important issue to investigate.

1.6.4 Cross Layer Optimization

The ambitious demands of massive connectivity, enhanced spectral efficiency, capacity, and energy efficiency in 5G requires further performance enhancement of NOMA. Cross layer optimization is an important technique to increase the NOMA system performance. The fairness constraints are important considerations in NOMA systems and hence issues of power allocation, user clustering and scheduling, and pilot are required to be jointly optimized in order to perform cross layer optimization in NOMA.

1.6.5 HARQ Design for NOMA

Hybrid automatic repeat request is an important component of the communication system, which is used to increase the transmission reliability.

In DL NOMA, the UE has two decoding blocks. The first decoding block is used to remove the intra-user interference and the second is to decode its own message after performing FEC. If decoding failure occurs, the UE requests retransmission from the BS. However, unlike conventional OMA, the retransmission at the BS poses a challenge due to superposition of NOMA protocol. In particular, TPA and pairing of retransmitted UE with other UEs at the BS, buffering of data from first transmission at the UE side, and when to apply HARQ combination are some of the HARQ design challenges for NOMA [48, 49]. Therefore, it is important to research for a sophisticated retransmission scheme and HARQ design for NOMA that will lower the number of retransmissions, increase throughput, and improve reliability.

1.6.6 Massive MIMO NOMA

Although MIMO techniques are applied and proposed for NOMA in the literature, NOMA-BF still faces many challenges, which include a problem of optimal joint user allocation and BF schemes, joint transmit, and BF and receiver complexity of the UE. In case of applying massive MIMO to NOMA where hundreds of antennas are coordinating to form a NOMA BF is not a simple task. In order to extend the application of MIMO and massive MIMO to more than two users, it is required to perform resource allocation in a multi-dimensional space, which is both analytically and computationally demanding [24].

1.6.7 Full Duplex NOMA

Recent developments in FD invalidate a long-held assumption in wireless communications that the radios can only operate in half duplex mode [50]. FD radios aim to double the spectrum efficiency by allowing simultaneous transmission and reception over the same frequency channel. However, the performance of FD radios is limited by SI cancellation and CCI [51]. Combining NOMA with FD communication can further enhance the spectrum efficiency and throughput. But combining NOMA with FD will introduce intra-user interference due to the superposition. This will increase the receiver complexity. The resource and power allocation, efficient joint SI and intra-user interference cancellation algorithm, and low-complexity receiver designs are some of the design challenges related to FD NOMA and require further investigations.

1.7 Conclusion

In this chapter, we reviewed the existing DL NOMA with SIC and highlighted several limitations related to SIC that can consequently degrade the performance of NOMA scheme in DL. In order to alleviate the problems posed by SIC, we proposed a PIC-based receiver structure for DL NOMA. The numerical results prove the superiority of proposed PIC-based receiver over SIC and hence it is a promising receiver choice for UE in 5G DL NOMA. However, this performance gain of PIC is achieved at the cost of higher computational complexity as compared to SIC.

In order to make the proposed PIC-based receiver a more suitable and promising choice for 5G DL NOMA, one possible area of future work can be focused on reducing its computational complexity by attempting to achieve a good trade-off between receiver performance and complexity. Other possible directions for future research may include investigating MIMO DL NOMA with a PIC-based UE receiver, integrating power control in the DL NOMA system with a PIC receiver, and proposing a novel design of decoder block at the UE receiver to reduce the multi-user interference.

References

[1] Andrews, J. G. S., Choi, B. W., Hanly, S. V., Lozano, A., Soong, C. K., and Zhang, J. C. (2014). What will 5G be?, *IEEE J. Select. Areas Commun.* 32, 1065–1082.

[2] Akylidiz, I. F., Nie, S., Lin, S.-C., and Chandrasekaran, M. (2016). 5G roadmap: 10 key enabling technologies. *Comput. Netw.* 106, 17–48.

[3] Xiaoming, D., Shanzhi, C., Shaohui, S., et al. (2014). "Successive interference cancelation amenable multiple access (SAMA) for future wireless communications," in *Proceedings of the IEEE International Conference on Communication Systems (ICCS),* Shenzhen, 222–226.

[4] Zheng, I. M. A., Quan, Z. Z., Guo, D. Z., et al. (2015). Key techniques for 5G wireless communications: network architecture, physical layer, and MAC layer perspectives. *Sci. China Inf. Soc.* 58, 1–20.

[5] Timotheou, S., and Krikidis, I. (2015). Fairness for non-orthogonal multiple access in 5G systems. *IEEE Signal Process. Lett.* 22, 1647–1651.

[6] Saito, Y., Benjebbour, A., Kishiyama, Y., and Nakamura, T. (2013). "System-level performance evaluation of downlink non-orthogonal multiple access (NOMA)," in *Proceedings of the IEEE 24th International*

Symposium on Personal Indoor and Mobile Radio Communications (PIMRC), London, 611–615.

[7] Nonaka, N., Kishiyama, Y., and Higuchi, K. (2014). "Non-orthogonal multiple access using intra-beam superposition coding and SIC in base station cooperative MIMO cellular downlink," in *Proceedings of the IEEE 80th Vehicular Technology Conference (VTC Fall)*, Vancouver, BC, 14–17.

[8] Yang, L., Benjebboiu, A., Xiaohang, C., Anxin, L., and Huiling, J. (2014). "Considerations on downlink non-orthogonal multiple access (NOMA) combined with closed-loop SU-MIMO," in *Proceedings of the 8th International Conference on Signal Processing and Communication Systems (ICSPCS)*, Rowville, VIC, 15–17.

[9] Nonaka, N., Benjebbour, A., and Higuchi, K. (2014). "System-level throughput of NOMA using intra-beam superposition coding and SIC in MIMO downlink when channel estimation error exists," in *Proceedings of the IEEE International Conference on Communication Systems (ICCS)*, Macau, 202–206.

[10] Benjebbour, A., Saito, Y., Kishiyama, Y., Anxin, L., Harada, A., and Nakamura, T. (2013). "Concept and practical considerations of non-orthogonal multiple access (NOMA) for future radio access," in *Proceedings of the International Symposium on Intelligent Signal Processing and Communications Systems (ISPACS)*, Rowville, VIC, 770–774.

[11] Dai, L., Wang, B., Yuan, Y., Han, S., and Wang, Z. (2015). "Non-orthogonal multiple access for 5G: solutions, challenges, opportunities, and future research trends," in *Proceedings of the IEEE Communications Magazine*, 53, 74–81.

[12] Hoshyar, R., Wathan, F. P., and Tafazolli, R. (2008). "Novel low-density signature for synchronous CDMA systems over AWGN channel," in *Proceedings of the IEEE Transactions on Signal Processing*, London, 1616–1626.

[13] Kschischang, F. R., Frey, B. J., and Loeliger, H. A. (2001). "Factor graphs and the sum-product algorithm," in *Proceedings of the* IEEE *Transactions on Information Theory*, New York, NY, 498–519.

[14] Al-Imari, M., Xiao, P., Imran, M. A., and Tafazolli, R. (2014). "Uplink non-orthogonal multiple access for 5G wireless networks," in *Proceedings of the 2014 11th International Symposium on Wireless Communications Systems (ISWCS)*, 781–785.

[15] Hoshyar, R., Razavi, R., and Al-Imari, M., (2010). "LDS-OFDM an efficient multiple access technique," in *Proceedings of the* IEEE *71st Vehicular Technology Conference*, 1–5.

[16] Nikopour, H., and Baligh, H. (2013). "Sparse code multiple access," in *Proceedings of the IEEE 24th Annual International Symposium on Personal, Indoor, and Mobile Radio Communications (PIMRC)*, London, 332–336.

[17] Zhang, S., Xu, X., Lu, L., Wu, Y., He, G., and Chen, Y. (2014). "Sparse code multiple access: an energy efficient uplink approach for 5G wireless systems," in *Proceedings of the IEEE Global Communications Conference*, Austin, TX, 4782–4787.

[18] Wu, Y., Zhang, S., and Chen, Y. (2015). "Iterative multiuser receiver in sparse code multiple access systems," in *Proceedings of the IEEE International Conference on Communications (ICC)*, London, 2918–2923.

[19] Yuan, Z., Yu, G., and Li, W. (2015). Multi-user shared access for 5G. *Telecommun. Netw.* 5, 28–30.

[20] Yuan, Z., Yu, G., Li, W., Yuan, Y., Wang, X., and Xu, J. (2016). "Multi-user shared access for internet of things," in *Proceedings of the IEEE 83rd Vehicular Technology Conference (VTC Spring)*, Nanjing, 1–5.

[21] Tao, Y., Liu, L., Liu, S., and Zhang, Z. (2015). A survey: several technologies of non-orthogonal transmission for 5G. *China Commun.* 12, 1–15.

[22] Tachikawa, S., and Marubayashi, G. (1987). Spread time spread spectrum communication systems. *IEEE Glob. Telecommun. Conf.* 165, 615–619.

[23] Kusume, K., Bauch, G., and Utschick, W. (2012). "IDMA vs. CDMA: analysis and comparison of two multiple access schemes," in *Proceedings of the IEEE Transactions on Wireless Communications*, New York, NY, 78–87.

[24] Ding, Z. et al. (2017). "Application of Non-Orthogonal Multiple Access in LTE and 5G Networks," in *Proceedings of the IEEE Communications Magazine*, 185–191.

[25] NTT Docomo (2012). "Requirements, Candidate Solutions and Technology Roadmap for LTE Rel-12 Onward," in *Proceedings of the RWS-120010, 3GPP Workshop on Release 12 Onward Ljubljana*, Slovenia, 11–12.

[26] Tomida, S., and Higuchi, K. (2011). "Non-orthogonal Access with SIC in Cellular Downlink for User Fairness Enhancement," in *Proceedings*

of the International Symposium on Intelligent Signal Processing and Communications Systems (ISPACS), Nusa Dua, 1–6.

[27] Takeda, T., and Higuchi, K. (2011). "Enhanced user fairness using non-orthogonal access with SIC in cellular uplink," in *Proceedings of the IEEE Vehicular Technology Conference*, Yokohama, 1–5.

[28] Xiaohang, C., Benjebbour, A., Yang, L., Anxin, L., and Huiling, J. (2014). "Impact of rank optimization on downlink non-orthogonal multiple access (NOMA) with SU-MIMO," in *Proceedings of the IEEE International Conference on Communication Systems (ICCS)*, Shenzhen, 233–237.

[29] Saito, Y., Kishiyama, Y., Benjebbour, A., Nakamura, T., Anxin, L., and Higuchi, K. (2013). "Non-orthogonal multiple access (NOMA) for cellular future radio access," in *Proceedings of the IEEE 77th Vehicular Technology Conference (VTC Spring)*, Nanjing, 2–5.

[30] Benjebbour, A., Anxin, L., Kishiyama, Y. et al. (2014). "System-level performance of downlink NOMA combined with SU-MIMO for future LTE enhancements," in *Proceedings of the 2014 Globecom Workshops (GC Wkshps)*, Austin, TX, 706–710.

[31] Ding, Z., Yang, Z., Fan, P., and Poor, H. V. (2014). On the performance of non-orthogonal multiple access in 5G systems with randomly deployed users. *IEEE Signal Process. Lett.* 21, 1501–1505.

[32] Buehrer, R. M. (2006). *Code Division Multiple Access (CDMA)*. San Rafael, CA: Morgan and Claypool.

[33] Ding, Z., Fan, P., and Poor, H. V. (2016). Impact of User Pairing on 5G Non-Orthogonal Multiple Access. *IEEE Trans. Vehic. Technol.* 65, 6010–6023.

[34] Buehrer, R. M., Correal-Mendoza, N. S., and Woerner, B. D. (2000). A simulation comparison of multiuser receivers for cellular CDMA. *IEEE Trans. Vehic. Technol.* 2000, 1065–1085.

[35] Zhao, R., Si, Z., He, Z., Niu, K., and Tian, B. (2014). "A joint detection based on the DS evidence theory for multi-user superposition modulation," in *Proceedings of the 4th IEEE International Conference on Network Infrastructure and Digital Content (IC-NIDC)*, Beijing, 390–393.

[36] Higuchi, K., and Kishiyama, Y. (2013). "Non-Orthogonal Access with Random Beam forming and Intra-Beam SIC for Cellular MIMO Downlink," in *Proceedings of the IEEE 78th in Vehicular Technology Conference (VTC Fall)*, New York, NY, 2–5.

[37] Ding, Z., Adachi, F., and Poor, H. V. (2016). "The application of MIMO to non-orthogonal multiple access," in *Proceedings of the EEE Transactions on Wireless Communications*, New York, NY, 537–552.

[38] Zhang, D., Zhou, Z., and Sato, T. (2015). Towards SE and EE in 5G with NOMA and Massive MIMO Technologies. *arXiv* 1504, 02212v1.

[39] Ding, Z., and Poor, H. V. (2016). "Design of massive-MIMO-NOMA with limited feedback," in *Proceedings of the IEEE Signal Processing Letters*, New York, NY, 629–633.

[40] Ding, Z., Peng, M., and Poor, H. V. (2015). Cooperative non-orthogonal multiple access in 5G systems. *IEEE Commun. Lett.* 19, 1462–1465.

[41] Men, J., and Ge, J. (2015). Non-orthogonal multiple access for multiple-antenna relaying networks. *IEEE Commun. Lett.* 19, 1686–1689.

[42] Kim, J. B., and Lee, I. H. (2015). Non-orthogonal multiple access in coordinated direct and relay transmission. *IEEE Commun. Lett.* 19, 2037–2040.

[43] Liu, Y., Ding, Z., Elkashlan, M., and Poor, H. V. (2016). Cooperative non-orthogonal multiple access with simultaneous wireless information and power transfer. *IEEE J. Select. Areas Commun.* 34, 938–953.

[44] Divsalar, D., and Simon, M. (1995). *Improved CDMA Performance Using Parallel Interference Cancellation*. Pasadena, CA: JPL Publication, 1995.

[45] Busching, F., Schildt, S. S. and Wolf, L. (2012). "DroidCluster: towards smartphone cluster computing – the streets are paved with potential computer clusters," in *Proceedings of the 32nd International Conference on Distributed Computing Systems Workshops (ICDCSW)*, New York, NY, 114–117.

[46] Qualcomm (2013). *Qualcomm Snapdragon Benchmark Report*. Available at: https://www.qualcomm.com/documents/qualcomm-snapdragon-benchmark-report

[47] Samsung Galaxy S4 Exynos 5 Octa Benchmarks (2016). Available at: http://www.fonearena.com/blog/67821/samsung-galaxy-s4-exynos-5-octa-benchmarks.html

[48] Li, A., Benjebbour, A., Chen, X., Jiang, H., and Kayama, H. (2015). "Investigation on hybrid automatic repeat request (HARQ) design for NOMA with SU-MIMO," in *Proceedings of the IEEE 26th Annual International Symposium on Personal, Indoor, and Mobile Radio Communications (PIMRC)*, Hong Kong, 590–594.

[49] Li, A., and Jiang, H. (2016). Method and apparatus for re-transmitting data. U.S. Patent No. 0 269 933.

[50] Bharadia, D., McMilin, E., and Katti, S. (2013). "Full duplex radios," in *Proceedings of the SIGCOMM Computer Communication*, New York, NY.

[51] Sun, Y., Ng, D. W. K., Ding, Z. and Schober, R. (2017). Optimal joint power and subcarrier allocation for full-duplex multicarrier non-orthogonal multiple access systems. *arXiv* 1607, 02668.

2

Beam Steering MIMO Antenna for Mobile Phone of 5G Cellular Communications Operating at MM-Wave Frequencies: Design

T. Thomas[1], Peter Gardner[1], Alexandros Feresidis[1] and K. Veeraswamy[2]

[1]Department of Electronic Electrical and Systems Engineering, School of Engineering, College of Engineering and Physical Sciences, University of Birmingham, Birmingham, United Kingdom
[2]QIS College of Engineering and Technology, Ongole, India

Abstract

Fifth-generation technology in connection with cellular mobile communication is aiming to provide higher data rates expecting up to tens of GBPS to the individual end user. This in return demands higher bandwidth allocation for every receiving device. The unused spectrum in the millimeter-wave range makes it possible, if both transmitting and receiving devices are upgraded to new frequency bands. In this context, the design of a MIMO antenna for mobile phones operating at 15 and 28 GHz is presented in this paper. As a substitute for a traditional Omni-directional antenna, high propagation losses of millimeter-wave frequencies stipulate the beam streaming phenomena with directional antennas in the mobile phone. The proposed 120×40 mm^2 RT/duroid 6002 substrate with a compact MIMO antenna design in 40×20 mm^2 is optimized for portable devices. Incorporation of an inductive switch gives multi-band operation. Compatibility to existing frequency bands namely LTE-700, GSM-850/GSM-900, and LTE-2200 is a noticeable aspect in this design. A gain of 7–8 dBi is achieved, which is expected to increase further for more directive operation. Eighty to ninety percent of efficiency is attained in this design, which emphasizes the good radiation capabilities of the proposed antenna. CST Microwave Studio-simulated results are presented.

2.1 Introduction

Fifth generation technology with high expectations in the commercial mobile communication sector is a topic of concern in view of both end users and network operators. The foreseen capabilities of 5G technology have opened many possibilities in mobile services and also set many design challenges. For the manifestation of 5G's credibility over preceding generations, higher bandwidth per user will be a key requirement, typically 1 Gbps. This enables a mobile user to have services like hologram and ultra-high definition. This in turn demands a 100-Gbps cell capacity in order to support Giga-bit mobile service. Other objectives in addition to the above are inter-network collaboration and seamless connectivity.

New mobile communication technology namely 5G, with its new proposed services and objectives, brings forth many design challenges. In spite of advancements in wireless cellular communication technology, scarcity of bandwidth for new high bandwidth applications drive the search for alternative methods and techniques. To make data transmit with high bit rate between two end terminals of the cellular communication system, many changes need to be made both at the transmitter and at the receiver end.

The aforementioned situation is a result of frequency shift to new bands. In other words, introduction of new frequency bands raises difficulties of using current installations. The tremendous growth of data usage both by public and by private sectors for commercial and entertainment purposes created higher data traffic, due to which new services can't be provided with current bands. The only alternative is to make use of unused spectrum that has been allotted for cellular mobile communication. MM-wave frequency bands provide alternatives to meet the bandwidth requirements.

Millimeter-wave-frequency signals are subject to high path loss. The preferable solution to use MM-Wave signals is to employee the directional antennas or focused beams toward the target. At the transmitter if a directional array antenna operating at MM-Wave frequency is used in place of a single Omni-antenna operating at low frequency, maintaining the same distance between the transmitter and the receiver, then same signal strength can be observed at the receiver end. Beam forming and beam steering are some techniques useful in this context.

Active analog phase shifters are of prime interest in this regard, Analog beam forming network using I/Q mixers for K-band, 3-dB branch line coupler terminated by two parallel-resonant circuits for 2.4 GHz, wideband active phase-shifter (I/Q network and VGAs) for 60–80 GHz, analog quadrature baseband phase shifters (Op-Amps and VGAs) for 2.4 GHz, adaptive array steered by local phase shifters (AA-LPS) for 2.4 GHz, double-balanced mixer based analog phase shifters for s-band, and graphene-based analog phase shifter designs for THz frequencies are proposed in [1–7], respectively. All these are for analog beam steering but not with desired specifications; hence, wide scan angle beam steering analog phase shifter operating at millimeter-wave for portable devices is still a design challenge. This results in one of the proposed objectives, with wide continuous phase shift, low insertion loss, two-dimensional scanning, low power, and low profile as design metrics. With multiple antenna elements beam steering will be possible and much effective, which results in a need for a MIMO system.

Multiple-input multiple-output technology in this context is used for high bit rate and for beam steering. The higher the number of elements in MIMO, the higher the flexibility in operation. On the other hand, MIMO design needs to be optimized in size. The MIMO designs proposed in [8–13] operate at one or two frequencies of interest but were not yet optimized aiming at a portable device like a mobile phone. Moreover, very limited literature is available describing the combined operation of MIMO with other circuits like beam steering network and active circulator. Hence, the proposed MIMO system is aimed for portable devices, with low profile and operating at frequencies 15 GHz, 22 GHz, 28 GHz, 38 GHz, ... etc., below 100 GHz.

In fact we are expecting that new features like sensors for household appliances, smart micro grids for every house, every part of automobile, smart highways, and complete distance education, etc. To achieve trans-portable handset application, specific requirements including small-size, built-in multi-band, coexistence of a multi radio system and a MIMO system, operating frequency, bandwidth, polarization, manufacturing cost, and the possible interaction effects with the handset have to be taken into consideration when designing the antenna.

2.2 The 5th-Generation Cellular Mobile Communications

2.2.1 Design Issues at Base Station for 5G Cellular Mobile Communication System

There are many areas in which developments are being made to establish a commercial system to bring 5G into reality. Handfuls of proposals are made by different research groups as shown in Figure 2.1. From a base station

MOBILE CELLULAR COMMUNICATION
NETWORK

Heterogeneous networks	Radio access networks	Ultra-dense networks (UDN)	Hybrid 5G environments	Small cells	
Virtual networks	Software controlled networks	Electronic packet switching (EPS) networks	Content-centric networks	Energy efficiency of networks	Mobile edge networks (MeNs)
Architecture of millimeter wave (mmW) RF subsystems	Architecture of TDD cellular system	Medium access control (MAC)	Full-Duplex radios		Frequency quadrature amplitude modulation (FQAM)

Base station antennas	Massive multi-input multi-output antennas	Beam forming

Ka-Band front end amplifier scenarios Dual-band bandpass filter	RF-MEMS 2-state basic attenuator modules	Non-stationary mobility pattern in the cell

Accesses modulation schemes
Algorithm
Coding
Physical-layer network coding (PNC)
Proof of connection
Interference mitigation

Channel
New frequency/mm-wave frequency
Channel modeling at THz Band

Figure 2.1   Overview of 5G system base station development issues.

(BS) point of view, the difficulties involved in using millimeter-waves are addressed with a variety of network proposals like heterogeneous networks, radio access networks, ultra-dense networks, hybrid networks, small cells, virtual networks, software-controlled networks, electronic packet switching networks, content-centric networks, energy-efficient networks, mobile edge networks, and many more. In addition to this some of the new architectures namely MM-Wave RF subsystem, time division duplexing (TDD) cellular system, and medium access control have been proposed. On the other hand BS antennas with massive multi-input multi-output scheme and beam steering are interesting proposals for 5G. The circuit side of the 5G system is also gradually taking momentum; hence, there were new amplifiers, dual-band band-pass filters, and RF-MEMS 2-state basic attenuator module designs under development. Accesses modulation schemes, algorithms, coding, physical-layer network coding, proof of connection, interface mitigation, and channel modeling are at the analysis and design stage for the 5G system.

2.2.2 Design Issues at User Equipment for 5G Cellular Mobile Communication System

Mobile phone/user equipment (UE) design for 5G is very challenging due to the expected new operating frequency and services and application it is expected to support. As of now some of the antenna models namely planar

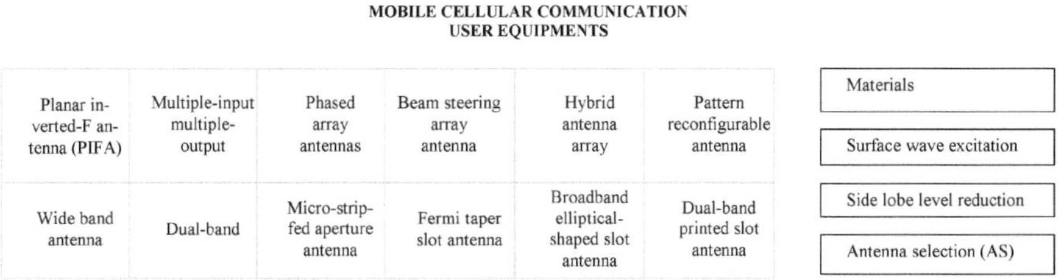

MOBILE CELLULAR COMMUNICATION
USER EQUIPMENTS

Planar in-verted-F antenna (PIFA)	Multiple-input multiple-output	Phased array antennas	Beam steering array antenna	Hybrid antenna array	Pattern reconfigurable antenna		Materials
							Surface wave excitation
Wide band antenna	Dual-band	Micro-strip-fed aperture antenna	Fermi taper slot antenna	Broadband elliptical-shaped slot antenna	Dual-band printed slot antenna		Side lobe level reduction
							Antenna selection (AS)

Figure 2.2   Overview of 5G system user equipment development issues.

inverted-F antenna (PIFA), MIMO, phased array antennas, beam steering array antennas, hybrid antenna array, pattern reconfigurable antennas, wide band antennas, dual-band micro strip-fed aperture antenna, Fermi taper slot antennas, broadband elliptical-shaped slot antennas, and dual-band printed slot antennas are trying to meet the design metrics of the 5G mobile phone. But improvement is yet to be achieved in the mobile phone design to make UE fully operative with all the services promised by 5G. Not all materials are suitable for design of mobile phone parts, particularly the antenna operating at MM-Wave frequencies. Hence, material characterization at MM-Wave frequencies has gained significance for research. Nature of excitation, adaptive beam forming, and steering do have considerable importance in view of 5G mobile phone design. Figure 2.2 shows the overview of the proposed design issues related to 5G mobile phone design.

2.2.3 Applications and Techniques Supported by 5G Technology

Many applications, such as GSM for railways (GSM-R), cellular network-based high-speed railway (HSR) wireless communication systems, are possible with the new capabilities of 5G networks. In view of application developers and device manufacturers, the 5G mobile networks bring a number of new possibilities for application developers. They cut down many barriers that have potentially prevented developing new applications for mobile networks by introducing significantly higher data rates, lower delays, and better energy efficiency. Wireless factory automation is an application area with highly demanding communication requirements. With the development of 5G mobile technology, the concept of dedicated radio-frequency (RF) charging promises to support the growing market of wearable devices as well. Smart grids, smart cities, and eHealth are other promising fields of interest which have many advantages with 5G abilities for the implementation of these services.

Emerging new applications, such as Internet of Things (IoT), gigabit wireless connectivity, tactile internet, and many more, are expected to impose new and diverse requirements on the design of the 5G of cellular communication systems. The IoT will facilitate a wide variety of applications in different domains, such as smart cities, smart grids, industrial automation (Industry 4.0), smart driving, assistance for the elderly, and home automation [14]. E-health and the IoT are two important applications that define the new requirements. Efforts to improve the potential of 5G networks and machine-to-machine (M2M) communications in a way that could help to develop and evolve mobile health (m-health) applications are still at preliminary stages.

The so-called 5G networks promise to be the foundations for the deployment of advanced services, conceived around the joint allocation and use of heterogeneous resources, including network, computing, and storage [15]. An optical-wireless 5G infrastructure offering converged fronthauling/backhauling functions to support both operational and end-user cloud services. 5G transport networks need to provide the required capacity, latency, and flexibility in order to integrate the different technology domains of radio, transport, and cloud. In addition, 5G cellular networks provide device-to-device real-time communications which can be used for real-time positioning. The Tactile Internet will be able to transport touch and actuation in real time. The primary application running over the Tactile Internet will be haptic communications. Design efforts for both the Tactile Internet and the haptic communications are at a nascent stage. It is expected that the next-generation (5G) wireless networks will play a key role in realizing the Tactile Internet. Communication technologies of the Tactile Internet have to achieve a combination of extremely low latency under high reliability and security constraints. The targeted applications are in the fields of industry automation and transport systems, health care, education, and gaming. 5G addresses tactile use cases under the term mission-critical machine type communication. The light fidelity (LiFi) technology is one of the promising solutions to increase transmission capacity in the indoor scenario. It is based on light emitting diodes (LEDs) to enable high-speed communication with fully networking capabilities. One issue in the implementation of LiFi is to select the appropriate access technique in the multiuser environment.

Recently, there has been increasing interest and rapid growth in MM-Wave antennas and devices for use in diverse applications, services, and technologies such as short-range communication, future MM-Wave mobile communication for the 5G cellular networks, and sensor and imaging systems. Millimeter-wave spectrum becomes the frontier for 5G cellular network due to the availability of massive bandwidth in 30–300 GHz radio spectrum which can be used for cellular communications and backhaul services. The MM-Wave frequencies (roughly above 10 GHz) offer the availability of massive bandwidth to increase the capacity of 5G cellular wireless systems. The available spectrum in the MM-Wave frequency band meets the spectrum scarcity in microwave frequencies and supply required bandwidth. Nevertheless, the use of these frequencies comes with challenges to the network and requires a new system design. However, to overcome the high isotropic pathloss at these frequencies, highly directional transmissions will be required at both the BS and the mobile UE to establish sufficient link budget in wide area networks.

The RF beam-steering network electronically steers the antenna beam in the desired direction by providing suitably phase-shifted signals to the array antenna elements. The traditional beam forming network design techniques can be mainly classified into four different typologies: RF phase shifting; LO (local oscillator) phase shifting; IF phase shifting; digital phased arrays. In the following, for the sake of simplicity, only the receiver section will be considered for discussion; however, similar issues are applicable to transmitter section as well [1].

- *RF phase shifting:* In this architecture, the signals of the antenna elements are phase shifted and combined at the RF level in such a way that the interfering signals coming from unwanted directions are filtered. Then, the combined signal is converted to baseband through a mixer. This is the most common and the most compact architecture for phased array antennas because it needs a single mixer and there is no need for distribution of the LO signal. The primary challenge is to find high-performance phase shifters which operate at RF with a reasonable cost. Regardless of the technology used for phase shifters, such as MEMS (Micro-electro-mechanical systems), LTCC (low-temperature cofired ceramic), GaAs, or CMOS, the passive phase shifters, in general, have high RF losses, whereas the active ones usually have a limited dynamic range. Another factor that must be considered is the amplitude variation of the signal across each channel, so the phase shifters also maintain a constant loss inside their phase shift range. In order to solve this problem, variable gain amplifiers can be used to level the signal amplitude on each channel. However, at the same time, these amplifiers should not introduce phase variations.

- *LO phase shifting:* In this architecture, the intermediate frequency (IF) signal phase is the difference between the RF signal phase and the LO signal phase on each channel. So, the phase adjustment of the IF signals is achieved by tuning the phase of the LO signal which excites each radiating element. Switchable time delays can be used to shift the phase of the 10 MHz reference signals in order to control the phase of the LO signals. In this case, the phase shifters are not located along the signal path. Then, the losses, the non-linearity, and the noise performances of the phase shifters do not have a direct impact on the system. However, this approach needs a large number of mixers and so, in general, is of a higher complexity [16].

- *IF phase shifting:* Also in this case, the IF signal on each channel is the difference between the RF signal phase and the LO signal. In this case the phase variation can be obtained through an accurate tuning of the IF signal phase. The characteristics of this architecture in terms of losses, non-linearity, and noise performances are usually better than the RF phase shifting technique, because the phase shifters are placed on the IF signal path, but, as for the LO phase shifting technique, also in this case the system needs a large number of mixers, thus increasing the overall complexity [1].

- *Digital phased arrays:* In the design of digital phased arrays, the signal at each radiating element is first digitized through an analog-to-digital converter (ADC) and then analyzed through a digital signal processing unit (DSP) [1]. In this architecture, the interfering signals coming from unwanted directions are not filtered by the analog section, but, instead, they are processed by the DSP unit. Therefore, all the elements, which elaborate the signals (such as the RF mixer, the ADC, etc.) before they arrive at the DSP, must have a sufficient dynamic range in order to manage also the interfering signals. The advantage of this architecture is its flexibility and the ability to simply create multiple beams. Also, the digital phased arrays can adaptively update their radiation pattern in order to monitor the desired signal through the main lobe of the beam and keep track of the interfering signals by inserting null values in their directions of arrival.

The phase shifting function (0–360°) is performed by the use of packaged I/Q mixers employed in a non-canonical configuration, achieving a vector modulation of the input RF signal. Indeed, by controlling the I/Q ports of the mixers with DC voltage signals, rather than with modulating signals as a rule, the resulting effect is to introduce a phase and gain variation to an arbitrary input RF signal.

Operating frequency bands of the LTE/GSM/UMTS and newly proposed 5G

- LTE 700 (698–787 MHz)
- LTE 2300 (2305–2400 MHz)
- LTE 2500 (2500–2690 MHz)
- GSM 850 (824–894 MHz)
- GSM 900 (880–960 MHz)
- GSM 1800 (1710–1880 MHz)
- GSM 1900 (1850–1990 MHz)
- UMTS band (1920–2170 MHz)
- 15 GHz
- 28 GHz
- 38 GHz.

2.3 Proposed Antenna: Design and Analysis

2.3.1 Proposed MIMO Antenna Model #1

The proposed MIMO antenna design is shown in Figure 2.3. The mobile system board is designed with an FR4 substrate having a permittivity of 4.4 and a loss tangent of 0.001. Dimensions suitable for smart phones are used in the design for the system board, i.e., 120×40 mm^2.

Figure 2.4 shows the dimensions of the antenna elements. Port 1 is joined to patch with the inductive element and port 2 is connected to the "U" patch. In the proposed design multiple antenna elements are employed to create the MIMO system, main antenna, and auxiliary antenna resonating at different frequencies are printed at the bottom edge of the system board which supports LTE-700, GSM-850, GSM-900, lower bands, and LTE2200. Moreover, 15 GHz and 28 GHz are also supported in the 10–30 GHz range [17].

The structure and the size decide a particular resonance frequency for the individual radiator elements. The employment of an inductive component overcomes the drawback of the patch antenna; this is one of the simplest methods to enhance the bandwidth at its resonant frequency of the patch antenna. The main patch antenna has 60-mm- and 70-mm-long strips with

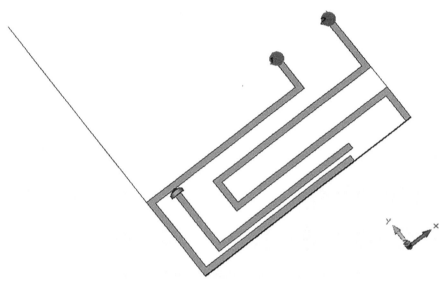

Figure 2.3   Proposed MIMO antenna model #1.

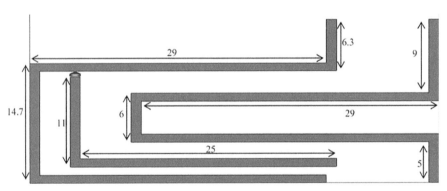

Figure 2.4   Dimensions of proposed model #1.

a 7.5 nH inductor suitably positioned to give the required resonance frequencies. Embedding inductive capacitive elements, making slits in the patch, and increasing the height of the substrate will enhance the bandwidth of the patch antenna. In the main antenna, an inductor is used to improve the bandwidth at resonant frequencies. Usage of thinner substrate will be appreciated in the design of slim modern smart phones. Higher antenna gain can be achieved by using thin substrates. An auxiliary antenna with a length of 75 mm is employed in the design. This part of the design is responsible for 2.2 GHz and lower LTE700 bands. As the length of the strip increases, the resonance frequency decreases. A quarter-wave length long strip is used to create resonance at required frequencies. The length of the strip is selected using the following equation.

$$L = \frac{C}{2F_r\sqrt{\varepsilon_{eff}}}, \tag{2.1}$$

where as in case of the rectangular patch antenna dimensions are calculated based upon the operating frequency. The dimensions of the radiator are calculated from the following equations.

$$W = \frac{1}{2F_r\sqrt{\varepsilon_o\mu_o}}\sqrt{\frac{2}{\varepsilon_r + 1}} \tag{2.2}$$

$$\varepsilon_{eff} = \frac{\varepsilon_r + 1}{2} + \frac{\varepsilon_r - 1}{2}\left[\frac{1}{1 + 12(H/W)} + 0.04\left(1 - \frac{W}{H}\right)^2\right] \tag{2.3}$$

$$L = \frac{C}{2F_r\sqrt{\varepsilon_{eff}}} - 2\Delta L \tag{2.4}$$

$$\Delta L = H \times 0.412\left(\frac{(\varepsilon_{eff} + 0.3)\left(\frac{W}{H} + 0.264\right)}{(\varepsilon_{eff} - 0.258)\left(\frac{W}{H} + 0.8\right)}\right) \tag{2.5}$$

Being an efficient method to enhance the data rate, the MIMO antenna system has much importance in 5G cellular communication. Nevertheless, insufficient space in the phone creates interference between the elements of MIMO. So, while implementing the MIMO system, the spacing between elements needs to be optimized so that it will give good isolation characteristics. Simplicity in structure and ease of fabrication make this proposed antenna system a considerable design for future mobile cellular communication handsets. Care has been taken in making the design compact to avoid the proximity effect of the adjacent components of the mobile phone.

In qualifying the proposed antenna: gain, efficiency, return loss, isolation, and far field radiation characteristics are used with the help of CST Microwave Studio. For smart phones, multi-band antennas are most appropriate. Ultra-wide band antennas are difficult to embed in mobile phones; hence, it is recommended that multi-band antenna systems be designed for the handset.

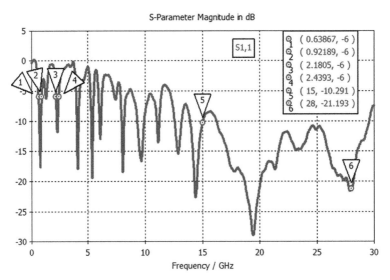

Figure 2.5   Return loss characteristics of proposed model #1: 0–30 GHz.

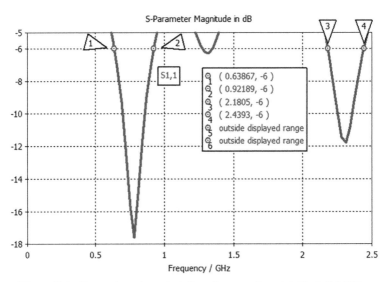

Figure 2.6   Return loss characteristics of proposed model #1: 0–2.5 GHz.

As shown in Figures 2.5 and 2.6, the resonance bands of the proposed antenna are 0.64–0.92 GHz and 2.18–2.43 GHz. Both the antenna elements exhibit good performance in the lower band. LTE700, GSM 850, GSM 900,

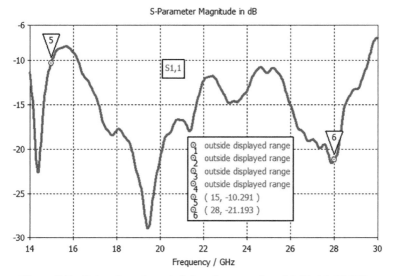

Figure 2.7 Return loss characteristics of proposed model #1: 14–30 GHz.

LTE2200, and few MM-Wave frequency bands are available in the range of the proposed antenna resonance band. In the upper 14–30 GHz band two MIMO antenna elements exhibit good resonance behavior particularly at 15 and 28 GHz frequency; refer to Figure 2.7.

The proposed antenna covers bands that are in usage by current cellular networks and the MM-Wave frequency bands supported by the proposed antenna available in the 9–60 GHz range have not yet come to commercial use by telecommunication network operators. In antenna design, physical dimensions will be chosen as per expected resonance frequency. To make the MIMO antenna system resonate at different operating frequencies, it is required to maintain elements with corresponding physical dimensions. All together give a multi-band antenna system.

Figure 2.8 shows isolation between two radiators of MIMO. It is obvious that the proposed system has good isolation characteristics, when both the antenna elements are kept apart. However, when the spacing between the elements is less, interference due to another element is significant. At required frequency bands the isolation is far smaller than −13 dB. In the lower band of resonance, isolation was poor which is expected to be enhanced by grounding techniques or spatial diversity techniques. In Figure 2.8 the interference due to another antenna is clearly evident form current distributions, which demonstrate the impact of one radiator on other at different frequencies.

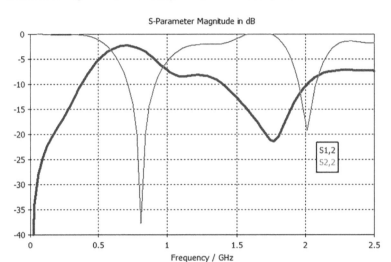

Figure 2.8   (A) Isolation characteristics of proposed model #1: 0–2.5 GHz.

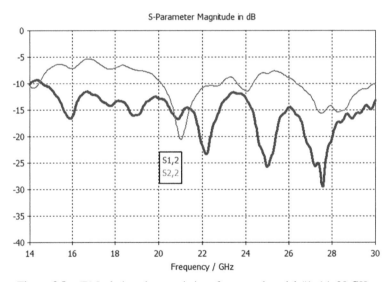

Figure 2.8   (B) Isolation characteristics of proposed model #1: 14–30 GHz.

At lower frequencies the isolation is poor, but at all other supporting upper frequencies it is well managed to avoid interference.

The proposed antenna has good radiation characteristics in the principal plane. Simulated far field 3D radiation characteristics are presented at

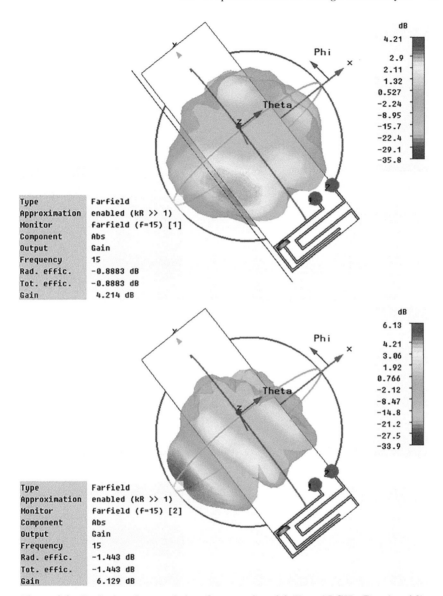

Type	Farfield
Approximation	enabled (kR >> 1)
Monitor	Farfield (f=15) [1]
Component	Abs
Output	Gain
Frequency	15
Rad. effic.	-0.8883 dB
Tot. effic.	-0.8883 dB
Gain	4.214 dB

Type	Farfield
Approximation	enabled (kR >> 1)
Monitor	Farfield (f=15) [2]
Component	Abs
Output	Gain
Frequency	15
Rad. effic.	-1.443 dB
Tot. effic.	-1.443 dB
Gain	6.129 dB

Figure 2.9 Radiation characteristics of proposed model #1: at 15 GHz (Port 1 and 2).

15 and 28 GHz in Figures 2.9 and 2.10. A gain value of 4–6.5 dBi at various resonance frequencies is obtained in this proposal; however, even higher gain values are necessary for some applications.

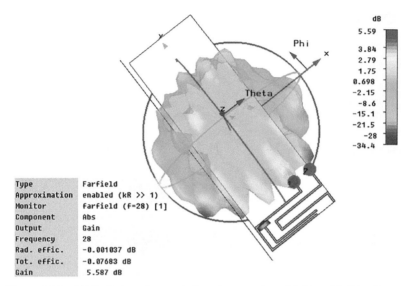

Type	Farfield
Approximation	enabled (kR >> 1)
Monitor	farfield (f=28) [1]
Component	Abs
Output	Gain
Frequency	28
Rad. effic.	-0.001037 dB
Tot. effic.	-0.07683 dB
Gain	5.587 dB

Figure 2.10   (A) Radiation characteristics of proposed model #1: at 28 GHz (Port 1).

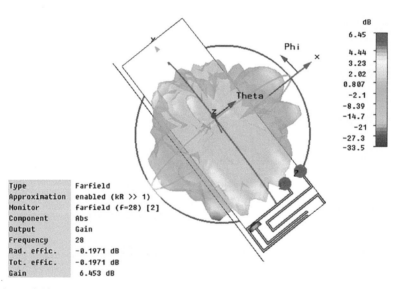

Type	Farfield
Approximation	enabled (kR >> 1)
Monitor	farfield (f=28) [2]
Component	Abs
Output	Gain
Frequency	28
Rad. effic.	-0.1971 dB
Tot. effic.	-0.1971 dB
Gain	6.453 dB

Figure 2.10   (B) Radiation characteristics of proposed model #1: at 28 GHz (Port 2).

2.3.2 Proposed MIMO Antenna Model #2

The proposed MIMO antenna model #2 is shown in Figure 2.11. The mobile system board is designed with an RT/duroid 6002 substrate having a permittivity of 2.94, a loss tangent of 0.001, and a thickness of 0.76 mm. Dimensions

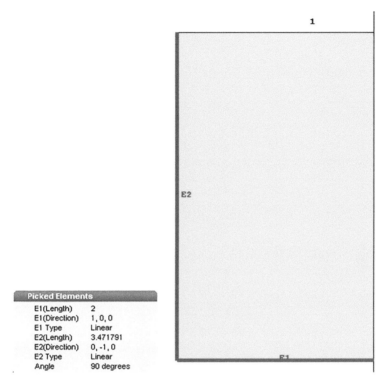

Picked Elements	
E1(Length)	2
E1(Direction)	1, 0, 0
E1 Type	Linear
E2(Length)	3.471791
E2(Direction)	0, -1, 0
E2 Type	Linear
Angle	90 degrees

Figure 2.11 Proposed model #2: Dimensions of radiator.

suitable for a smart phone are used in the design for the system board, i.e., 105×40 mm^2.

In the proposed model #2 only a single antenna element is employed resonating at different frequencies and is printed at the bottom edge of the system board which supports a wide band of ~20 GHz starting for 10 GHz, but the frequencies of interest are 15 and 28 GHz.

The thinner substrate is well appreciated in the design of slim modern smart phones to ensure a higher antenna gain. A radiator of size 2×3.47 mm^2 is employed in the design. The antenna impedance has been made approximately 50 Ω, for better impedance matching between the feed and the radiator. Commercially available subminiature version A (SMA) connectors have a 50 Ω port impedance; hence impedance matching yields good return loss characteristics. The resonance at 28 GHz can be achieved with the antenna dimensions obtained by using Equations (2.6–2.10), which are used to evaluate the dimensions of the radiator based on the substrate characteristics and the expected operating frequency. The design is responsible for resonating at 28- and 15-GHz bands. As the length of the strip increases, the resonance frequency decreases. Quarter-wave length long strips are used to create resonance at required frequencies. The length of the strip is selected using following equation.

$$L = \frac{C}{2F_r\sqrt{\varepsilon_{eff}}} \tag{2.6}$$

$$W = \frac{1}{2F_r\sqrt{\varepsilon_o\mu_o}}\sqrt{\frac{2}{\varepsilon_r+1}} \tag{2.7}$$

$$\varepsilon_{eff} = \frac{\varepsilon_r+1}{2} + \frac{\varepsilon_r-1}{2}\left[\frac{1}{1+12(H/W)} + 0.04\left(1-\frac{W}{H}\right)^2\right] \tag{2.8}$$

$$L = \frac{C}{2F_r\sqrt{\varepsilon_{eff}}} - 2\Delta L \tag{2.9}$$

$$\Delta L = H \times 0.412\left(\frac{(\varepsilon_{eff}+0.3)\left(\frac{W}{H}+0.264\right)}{(\varepsilon_{eff}-0.258)\left(\frac{W}{H}+0.8\right)}\right) \tag{2.10}$$

Width: 2 mm
Height: 0.76 mm
Width/Height: 2.632
Effective Dielectric Constant: 2.381
Impedance: 49.24 Ω
Length: 3.74 mm

As shown in Figure 2.12, the resonance band of the proposed antenna is 14–30 GHz. The radiating element covers the lower and upper parts of the band, namely 15 and 28 GHz. The multi-resonance results in a wide frequency band. The return loss at 28 GHz is -26.7 dB and at 15 GHz is -24.1 dB. The radiation pattern of the antenna at 15 and 28 GHz is shown in Figures 2.13 and 2.14. It is clear from the graph that the antenna has very good efficiency and gain values. And the current distribution of the antenna and the surface waves can be seen in Figure 2.15. The ground plane underneath the system board and the radiating element form a combination and the energy will be radiated in the direction somewhere between two edges of the two structures. As the underneath ground plane is lengthier than the radiating element, the main lobe will be near the corresponding edge of the ground plan. And the resonance occurs at different fractions of λ (wavelength) like $\lambda/2$, $\lambda/4$, $\lambda/8$, etc. This results in a very wide impedance bandwidth. But the

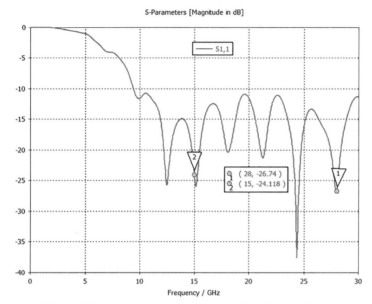

Figure 2.12 Proposed model #2: Return loss characteristics.

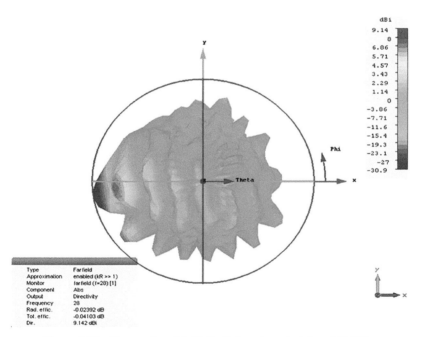

Figure 2.13 Proposed model #2: Radiation characteristics at 28 GHz.

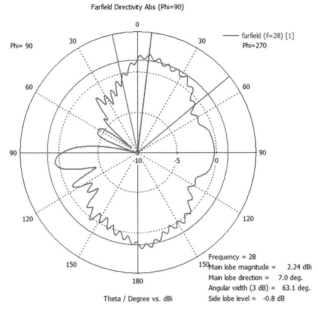

Figure 2.14   Proposed model #2: 2D radiation characteristics at 28 GHz.

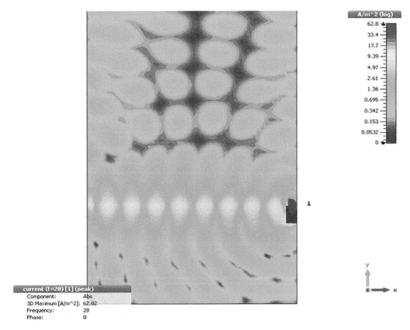

Figure 2.15   Proposed model #2: Current pattern at 28 GHz.

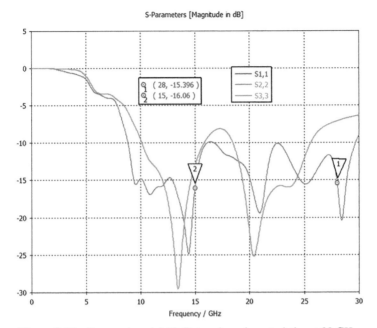

Figure 2.16 Proposed model #3: Return loss characteristics at 28 GHz.

isolation between elements will be poor if multiple elements are used to form a MIMO array antenna. Other techniques like split, spatial diversity, etc. are to be used to overcome this drawback.

The return loss characteristics of the proposed antenna array show that the antenna operates well at 15 and 28 GHz. Moreover, the antenna is resonating at multiple wide bands, having a bandwidth greater than 10 GHz as shown in Figure 2.16.

2.3.3 Proposed MIMO Antenna Model #3

As shown in Figure 2.17 three radiating elements are used in this array. The idea is to make array suitable for beam steering. All elements are of similar type and with the same dimensions. The more the array elements, the more the convenience in getting the required pattern, but the isolation between elements, which depends on inter-element distance, is considerable limiting point for the number of array elements. At the initial stages of this work it is reasonable to include at least three elements in this array. Thus, there is a poor isolation among these three elements.

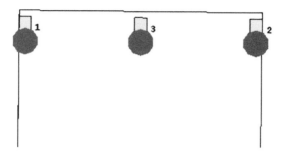

Figure 2.17　Proposed model #3: 3-element antenna array.

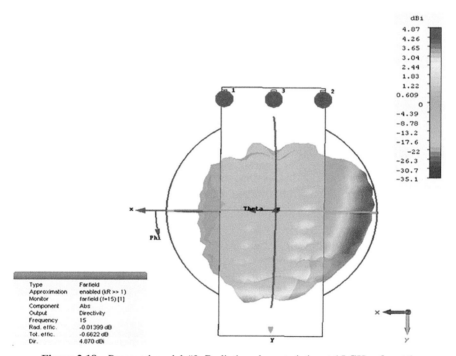

Figure 2.18　Proposed model #3: Radiation characteristics at 15 GHz of port 1.

The radiation characteristics of the proposed antenna array have been presented in Figures 2.18–2.23. Three individual radiation elements and their respective efficiencies and directive gain are shown in the figures.

In most of the cases, the efficiency of the radiators is near 90% and the gain of individual elements is in the range of 3.5–5.8 dBi. The radiation patterns are simulated at 15 and 28 GHz.

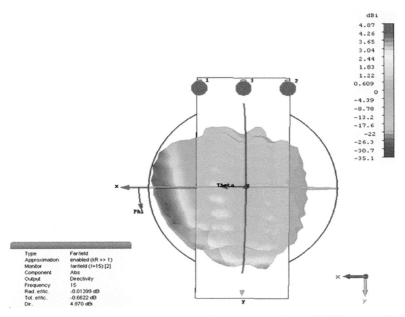

Figure 2.19 Proposed model #3: Radiation characteristics at 15 GHz of port 2.

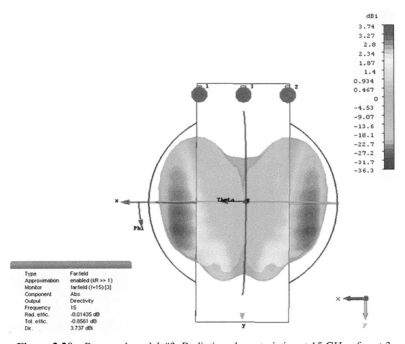

Figure 2.20 Proposed model #3: Radiation characteristics at 15 GHz of port 3.

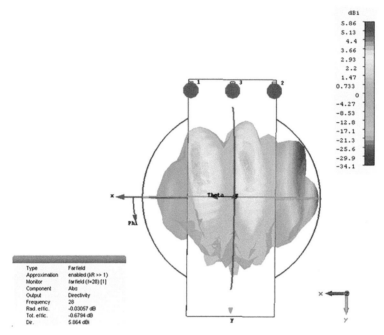

Figure 2.21   Proposed model #3: Radiation characteristics at 28 GHz of port 1.

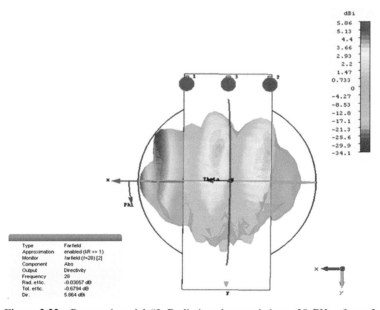

Figure 2.22   Proposed model #3: Radiation characteristics at 28 GHz of port 2.

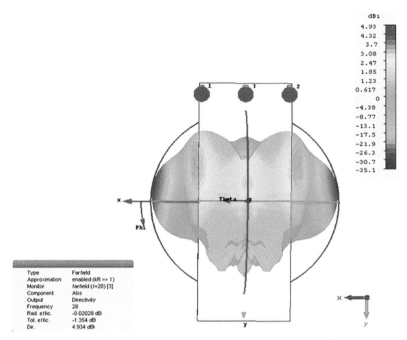

Figure 2.23 Proposed model #3: Radiation characteristics at 28 GHz of port 3.

The inherent propagation losses of MM-Wave frequency signals can be overwhelmed by using directional properties of the antenna. So it is strongly believed that both the transmitter and receiver should be directional and pointing to each other to get good signal at the receiving end. In case of mobile phones, small portable devices, achieving beam-steering is a challenging task. The proposed model is an attempt to achieve beam steering at MM-Wave frequencies. Figures 2.24–2.27 illustrate how the direction of the major lobe (high percentage of energy flow) of the antenna can be altered by applying different phase combinations to elements of the array.

2.4 Conclusion

The proposed models for the MIMO antenna system suitable for 5G mobile communication have good radiation and return loss characteristics. All models are designed to operate at desired frequencies namely 15 and 28 GHz. Particularly, model 1 has some lower frequency bands like LTE700, GSM 850, GSM 900, and LTE2200. The attained gains are in the range

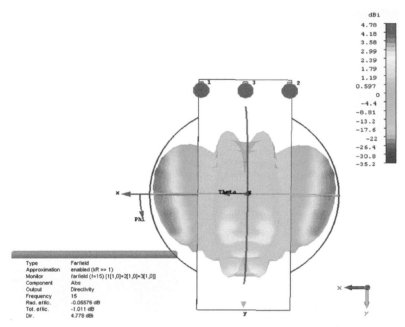

Figure 2.24　Proposed model #3: Beam steered pattern at 15 GHz 0° phase shift.

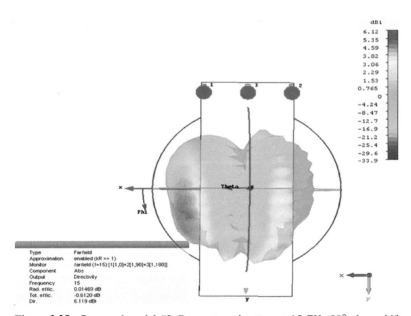

Figure 2.25　Proposed model #3: Beam steered pattern at 15 GHz 90° phase shift.

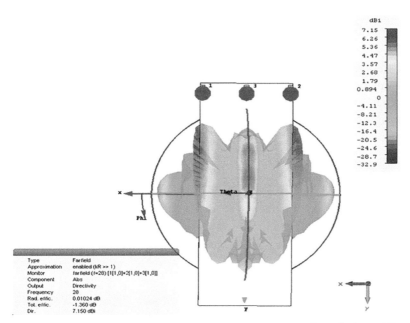

Figure 2.26 Proposed Model #3: Beam steered pattern at 28 GHz 0° phase shift.

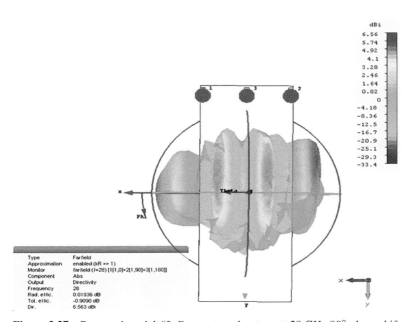

Figure 2.27 Proposed model #3: Beam steered pattern at 28 GHz 90° phase shift.

of approximately 3–6 dBi. Model 2 has very good radiation and efficiency characteristics. The wide operating bandwidth of this design is another impressive issue. Beam steering is attained by exciting three elements with different combinations of phase values. Analog beam steering is always an ideal feature but the presented proposal is yet to be improved to obtain more accurate beam forming and steering. Apart from the advantages, poor isolation among three elements is yet a challenging issue to be considered for further work.

Acknowledgments

The authors would like to thank the University of Birmingham for the facilities provided to complete the work.

References

[1] Alessandro, S., De Bilio, M. C., Pomona, I., Coco, S., Bavetta, G., and Laudani, A. (2015). "Analog beamforming network for Ka band satellite on the move terminal with phase shifting technique based on I/Q mixer," in *Proceedings of the European Radar Conference (EuRAD)* (Paris: IEEE), 445–448.

[2] Hung, W. H., Lam, H., Fung Y. W., Wei, D. C., and Wu, K.-L. (2011). "An active RFID indoor positioning system using analog phased array antennas," in *Proceedings of the Asia-Pacific Microwave Conference* (Melbourne, VIC: IEEE), 179–182.

[3] Kim, S. Y., Kang, D.-W., Koh, K.-J., and Rebeiz, G. M. (2012). An improved wideband all-pass I/Q network for millimeter-wave phase shifters. *IEEE Trans. Microw. Theory Tech.* 60, 3431–3439.

[4] Aerts, W., Delmotte, P., and Vandenbosch, G. A. E. (2009). Conceptual study of analog baseband beam forming: design and measurement of an eight-by-eight phased array. *IEEE Trans. Antennas Propag.* 57, 1667–1672.

[5] Kasami, H., Itoh, K., Shibata, O., Shoki, H., and Obayashi, S. (2002). Periodical intermittent interference suppression algorithm for 2.4-GHz-band adaptive array. *IEEE Veh. Technol. Conf.* 1, 435–439.

[6] Miura, A., Fujino, Y., Taira, S., Obara, N., Tanaka, M., Ojima, T., and Sakauchi, K. (2005). *S*-band active phased array antenna with analog phase shifters using double-balanced mixers for mobile SATCOM vehicles. *IEEE Trans. Antennas Propag.* 53, 2533–2541.

[7] Chen, P.-Y., Argyropoulos, C., and Alù, A. (2013). Terahertz antenna phase shifters using integrally-gated graphene transmission-lines. *IEEE Trans. Antennas Propag.* 61, 1528–1527.

[8] Halvarsson, B., Karam, E., Nyström, M., Pirinen, R., Simonsson, A., Zhang, Q., et al. (2016). "Distributed MIMO demonstrated with 5G radio access prototype," *Proceedings of the 2016 European Conference on Networks and Communications (EuCNC),* (Athens: IEEE), 302–306.

[9] Ali, M. M. M., and Sebak, A.-R. (2016). "Design of compact millimeter wave massive MIMO dual-band (28/38 GHz) antenna array for future 5G communication systems" in *Proceedings of the 17th International Symposium on Antenna Technology and Applied Electromagnetics (ANTEM)* (IEEE: Montreal, QC), 1–2.

[10] Li, Y., Wang, C., Yuan, H., Liu, N., Zhao, H., and Li, X. (2016). A 5G MIMO antenna manufactured by 3D printing method. *IEEE Antennas Wirel. Propag. Lett.* pp. 1–1.

[11] Guan, L., Rulikowski, P., and Kearney, R. (2016). Flexible practical multi-band large scale antenna system architecture for 5G wireless networks. *Electron. Lett.* 52, 970–972.

[12] Qin, Z., Geyi, W., Zhang, M., and Wang, J. (2016). Printed eight-element MIMO system for compact and thin 5G mobile handest. *Electron. Lett.* 52, 416–418.

[13] Agrawal, A., and Natarajan, A. (2016). "2.2 a scalable 28GHz coupled-PLL in 65nm CMOS with single-wire synchronization for large-scale 5G mm-wave arrays," in *Proceedings of the 2016 IEEE International Solid-State Circuits Conference (ISSCC),* San Francisco, CA, 38–39.

[14] Weitkemper, P., Koppenborg, J., Bazzi, J., Rheinschmitt, R., Kusume, K., Samardzija, D., et al. (2016). "Hardware experiments on multi-carrier waveforms for 5G," in *Proceedings of the IEEE Wireless Communications and Networking Conference,* San Francisco, CA, 1–6.

[15] Casellas, R., Muñoz, R., Vilalta, R., and Martínez, R. (2016). "Orchestration of IT/cloud and networks: from Inter-DC interconnection to

SDN/NFV 5G services," in *Proceedings of the International Conference on Optical Network Design and Modeling (ONDM))*, Cartagena, 1–6.

[16] Nistal-González, I., Otto, S., Litschke, O., Bettray, A., Wunderlich, L., Gieron, R., et al. (2014). "Planar phased array antenna for nomadic satellite communication in Ka-Band," in *Proceedings of the 11th European Radar Conference*, (Rome: IEEE), 396–399.

[17] Thomas, T., Charishma, G., and Veeraswamy, K. (2015). "MIMO antenna system with high gain and low SAR at for UE of 5G operating MM wave: design," in *Proceedings of the 10th International Conference on Information, Communications and Signal Processing (ICICS)* (Singapore: IEEE), 1–5.

3

Random Linear Network Coding with Source Precoding for Multi-session Networks

Xiaoli Xu[1], Yong Zeng[2] and Yong L. Guan[1]

[1] Nanyang Technological University, Singapore, Singapore
[2] National University of Singapore, Singapore, Singapore

Abstract

While random linear network coding (RLNC) asymptotically achieves the capacity of single-session multicast networks, it is sub-optimal in general for multi-session networks, since with RLNC, the inter-session interference is mixed with the desired information to the largest extend. In general, finding the optimal network code for multi-session networks is challenging. In this chapter, we show that effective network codes can be constructed for some classes of multi-session networks by applying RLNC at the intermediate nodes, together with properly designed linear precoding at the source nodes that minimize the inter-session interference. Specifically, we design the optimal precoding scheme and derive the achievable rate region for double-unicast networks with RLNC applied at the intermediate nodes. We also show that the capacity of multi-source single-sink erasure networks can be asymptotically achieved by RLNC with random linear precoder over a sufficiently large number of time extensions.

3.1 Introduction

In traditional communication networks, information packets are usually treated as physical commodities which can only be routed throughout the network without alteration. On the other hand, with network coding [1], the

65

packets are treated as information and allowed to be coded at the intermediate nodes. Compared with conventional routing, network coding has many promising advantages, such as the enhancement in network throughput, higher reliability, higher energy efficiency and better security. As a result, network coding has found a wide range of applications in ad hoc networks [2], wireless mesh networks [3], peer-to-peer communication systems [4] and distributed data storage systems [5].

Depending on the information demand, networks can be classified into two categories, namely *single-session networks* and *multi-session networks.* In single-session networks, all sink nodes demand identical information from a common source node. In contrast, for multi-session networks, different sink nodes may request information from different source nodes. It is well known that network coding can generally achieve higher throughput than routing by sending more information via fewer packet transmissions for both single- and multi-session networks. In particular, the simple linear network coding, where the network coded packets are generated by linearly combining the input packets, has been shown to be capacity-achieving for single-session multicast networks [6]. One important class of linear network coding is known as *random linear network coding* (RLNC), where the coding coefficients are randomly chosen from a finite field, and stored and updated in a coding vector attached with each packet. Compared with deterministic linear network coding schemes, RLNC is more robust for network dynamics due to its distributive nature of code construction. Besides, RLNC is very efficient for correcting packet erasures [7]. If the field size is sufficiently large, RLNC has been shown to achieve the capacity of single-session multicast networks with very high probability [8].

On the other hand, for multi-session networks, it remains unknown how to construct the optimal network code. In general, linear network coding is inadequate to achieve the multi-session network capacity, as evident from a multi-session network constructed in [9] based on Fano and non-Fano matroids. However, due to its low complexity, linear network coding is preferred in most practical scenarios. Thus, linear network coding schemes have received significant interests for multi-session networks. For example, a "poison-antidote" approach was proposed in [10] by searching for the butterfly structures in the network using linear optimization method. Besides, Wang et al. [11] introduced a pairwise inter-session network coding approach, where the coding operations are restricted to two symbols to reduce the inter-session interference. Recently, Heindlmaier et al. proposed an inter-session network coding by creating virtual multicast sessions, within which RLNC is applied [12].

All the aforementioned approaches for multi-session network code constructions are based on solving complicated optimization problems whose complexity grows quickly with the network size. To the authors' best knowledge, explicit construction of inter-session network coding for general multi-session networks has not been reported in the literature. However, if the numbers of sources and sinks are small, network code construction can be made much simpler. For instance, RLNC with properly designed source precoding has been shown to be optimal for networks with single source node and two sink nodes [13]. Besides, for double-unicast networks where two source-receiver pairs share common network resources, an achievable rate region can be obtained by the "rate-exchange" method [14], where starting from the single-user rate for one of the source-receiver pair, a non-zero rate for the other user is achieved by sacrificing the single-user rate via interference nulling. Furthermore, interference alignment was shown to achieve half of the min-cut for some three-unicast networks [15]. In this chapter, we aim to construct explicit linear network codes for some particular multi-session networks, including double-unicast networks [16] and multi-source erasure networks [17], by applying RLNC at the intermediate nodes and optimized precoding at the source node with the inter-session interference properly minimized.

Notations: Throughout this chapter, vectors and matrices are represented by boldface lower- and upper-case letters, respectively. For a matrix $\mathbf{A}$, we use $\mathbf{A}^T$, $\mathbf{A}^{-1}$, and $\mathrm{rank}(\mathbf{A})$ to denote its transpose, inverse, and rank, respectively. $\mathbf{0}_{n \times m}$ represents a zero matrix of size $n \times m$ and the subscripts are omitted when no ambiguity will be caused. The range (or column space) and null space of a matrix $\mathbf{A}$ are denoted by $\mathcal{R}(\mathbf{A})$ and $\mathcal{N}(\mathbf{A})$, respectively. $H(\cdot)$ is the entropy function, and $(x)^+$ is defined as $(x)^+ \triangleq \max\{0, x\}$.

3.2 Network Model with RLNC

As shown in Figure 3.1, we consider a multi-session network modeled by a directed acyclic graph $G = (V, E)$, which contains a set of source nodes, denoted by $S = \{s_1, ..., s_{|S|}\}$, and a set of sink nodes, denoted by $T = \{t_1, ..., t_{|T|}\}$. For $v \in V$, the edges entering into v and leaving from v are denoted by the sets $\mathrm{In}(v)$ and $\mathrm{Out}(v)$, respectively. Without loss of generality, we assume that there is no incoming edge to the source nodes and no outgoing edge from the sink nodes, i.e., $\mathrm{In}(s) = \mathrm{Out}(t) = \emptyset, \forall s \in S, t \in T$.

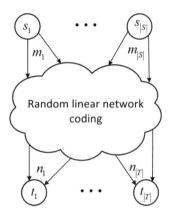

Figure 3.1   A multi-session multicast network with random linear network coding (RLNC).

We assume that each edge in the network is capable of carrying one symbol/packet per time slot.[1] By Merger's theorem, the minimum cut between the source subset $s_{N_1} \subseteq S$ and the sink subset $t_{N_2} \subseteq T$ equals the number of edge disjoint paths from s_{N_1} to t_{N_2}, denoted as $k_{N_1-N_2}$, where $N_1 \subseteq \{1, 2, ..., |S|\}$ and $N_2 \subseteq \{1, 2, ..., |T|\}$. It then follows that the number of outgoing edges from the source node s_i, denoted as m_i, is no smaller than $k_{i-\{1,..,|T|\}}$. Furthermore, if $m_i > k_{i-\{1,..,|T|\}}$, we may add an imaginary source s_i' connecting to s_i via $k_{i-\{1,..,|T|\}}$ edges without affecting the network capacity. Therefore, without loss of generality, we assume that $m_i = k_{i-\{1,..,|T|\}}$, i.e., there are $k_{i-\{1,..,|T|\}}$ outgoing edges from source s_i. Similarly, the number of incoming edges to sink t_j, denoted as n_j, is assumed to be equal to $k_{\{1,..,|S|\}-j}$ [18].

We assume that RLNC is applied at all intermediate nodes over finite field $\mathbb{F}_q$. The network coded symbols received by sink node t_j can then be expressed as:

$$\mathbf{y}_j = \sum_{i=1}^{|S|} \mathbf{H}_{ji}\mathbf{x}_i, \tag{3.1}$$

where the summation is performed over the finite field, $\mathbf{x}_i \in \mathbb{F}_q^{m_i}$ consists of the symbols sent over m_i outgoing edges of the source node s_i, and $\mathbf{H}_{ji} \in \mathbb{F}_q^{n_j \times m_i}$ represents the transition matrix from s_i to t_j, which is determined by the network topology and the network coding coefficients used at the intermediate nodes.

---

[1]Those edges with higher capacity can be modeled by multiple parallel edges.

It is observed that the input–output relationship given in Equation (3.1) is similar to that of the multiple-input multiple-output (MIMO) wireless communication systems. However, there are two main differences between them. First, different from the MIMO wireless communication systems, the additive noise terms are absent in Equation (3.1). This thus gives a *finite-field deterministic* communication model. Second, in contrast to wireless channels where the channel matrices are usually assumed to be generic and independent of each other, the transition matrices in Equation (3.1) are *correlated* due to network topology constraint. Hence, the existing results on wireless MIMO channel cannot be directly applied to the design of the precoding schemes for multi-session networks.

3.3 Precoder Design and Achievable Rate Region for Double-Unicast Networks

Multiple-unicast networks, in which each message is generated by one source node and demanded by exactly one sink node [19], is one class of multi-session networks of particular interest. It has been found that for any directed acyclic multi-session network, there exists a corresponding multiple-unicast network that has the same solvability over any alphabet [20]. As the simplest case of general multiple-unicast networks where network coding is relevant, double-unicast networks with two source-sink pairs have received significant research interests recently. For such networks, the achievability of the rate pair $(1, 1)$ has been established with various methods [21, 22]. On the other hand, since the capacity region for double-unicast networks remains unknown, many works have been devoted to deriving outer and inner bounds for the capacity region [14, 18, 23].

In this section, we study the achievable rate region for double-unicast networks by assuming that RLNC is employed at the intermediate nodes. Under such an assumption, the design problem reduces to finding the optimal strategies at the source and sink nodes so that the achievable rate region is maximized. Different from that in single-session networks, the optimal strategy at each source node for double-unicast networks needs to achieve a good balance between maximizing the rate to its corresponding sink node and minimizing the inter-session interference to the other. For a given set of network coding coefficients chosen by the intermediate nodes, such a scenario can be modeled as a deterministic interference channel (IC), whose capacity region has been obtained in [24]. For the particular class of *linear* deterministic IC of our interest, in which the outputs and the interferences are *linear* deterministic functions of the inputs, we show that the capacity region specified in [24] can be achieved by linear precoding and decoding at the source and sink nodes, respectively. The main techniques used in the proposed linear scheme are Han-Kobayashi rate splitting [25] and concatenated precoding composed of zero-forcing and random block-diagonal precoding matrices. Specifically, the information bearing symbols are first split into a common part which is decodable at both sink nodes, as well as a private part which is decodable at the designated sink node only. The resulting data symbols are then precoded by concatenated precoding matrices, which are carefully designed such that the interferences caused by the undesired data symbols are minimized at both sink nodes.

3.3.1 An Optimal Achievable Rate Region with RLNC

For double-unicast networks, the input–output relationship in Equation (3.1) reduces to

$$\begin{aligned}
\mathbf{y}_1 &= \mathbf{H}_{11}\mathbf{x}_1 + \mathbf{H}_{12}\mathbf{x}_2 \\
\mathbf{y}_2 &= \mathbf{H}_{21}\mathbf{x}_1 + \mathbf{H}_{22}\mathbf{x}_2,
\end{aligned} \tag{3.2}$$

It is observed that the input–output relation given in Equation (3.2) belongs to a special class of the deterministic IC considered in [24], in which the outputs $\mathbf{y}_1$ and $\mathbf{y}_2$ and the interferences $\mathbf{z}_1 = \mathbf{H}_{21}\mathbf{x}_1$ and $\mathbf{z}_2 = \mathbf{H}_{12}\mathbf{x}_2$ are *linear* deterministic functions of the inputs $\mathbf{x}_1$ and $\mathbf{x}_2$. By applying the result in [24], the capacity region of Equation (3.2) is given by the union of the set of all rate pairs (R_1, R_2) satisfying

$$R_1 \leq H(\mathbf{y}_1|\mathbf{z}_2) \tag{3.3}$$

$$R_2 \leq H(\mathbf{y}_2|\mathbf{z}_1) \tag{3.4}$$

$$R_1 + R_2 \leq H(\mathbf{y}_1) + H(\mathbf{y}_2|\mathbf{z}_1, \mathbf{z}_2) \tag{3.5}$$

$$R_1 + R_2 \leq H(\mathbf{y}_1|\mathbf{z}_1, \mathbf{z}_2) + H(\mathbf{y}_2) \tag{3.6}$$

$$R_1 + R_2 \leq H(\mathbf{y}_1|\mathbf{z}_1) + H(\mathbf{y}_2|\mathbf{z}_2) \tag{3.7}$$

$$2R_1 + R_2 \leq H(\mathbf{y}_1) + H(\mathbf{y}_1|\mathbf{z}_1, \mathbf{z}_2) + H(\mathbf{y}_2|\mathbf{z}_2) \tag{3.8}$$

$$R_1 + 2R_2 \leq H(\mathbf{y}_1|\mathbf{z}_1) + H(\mathbf{y}_2) + H(\mathbf{y}_2|\mathbf{z}_1, \mathbf{z}_2), \tag{3.9}$$

over all input distributions $\mathbf{x}_1 \in \mathbb{F}_q^{k_1-12}$ and $\mathbf{x}_2 \in \mathbb{F}_q^{k_2-12}$.

Note that the capacity-achieving scheme given in [24] is based on information-theoretic superposition coding, which is difficult to be implemented in practice. In the next section, we show that for the particular

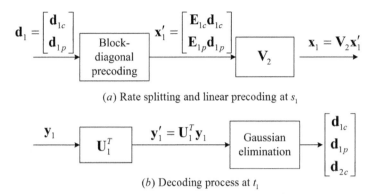

(*a*) Rate splitting and linear precoding at s_1

(*b*) Decoding process at t_1

Figure 3.2 Linear precoding and decoding at s_1 and t_1.

linear deterministic IC given by Equation (3.2), the capacity region specified in Equations (3.3–3.9) can be achieved with simple *linear* precoding and decoding at the source and the sink nodes, which thus provides a rate region achievable by linear coding schemes for double-unicast networks.

3.3.2 A Linear Capacity-achieving Scheme

Figure 3.2 provides a schematic overview of our proposed linear capacity-achieving scheme, which consists of rate-splitting, block-diagonal precoding, and invertible linear transformations at the source nodes, and invertible linear transformations and Gaussian eliminations at the sink nodes.

Before presenting the detailed designs for each block in Figure 3.2, we first reformulate the capacity region specified in Equations (3.3–3.9) in terms of the channel ranks based on the following lemma.

Lemma 1. *Let* $\mathbf{x}$ *be a random vector of dimension* $l \times 1$ *and* $\mathbf{A} \in \mathbb{F}_q^{p_1 \times l}$ *and* $\mathbf{B} \in \mathbb{F}_q^{p_2 \times l}$ *are two given matrices. We then have* $H(\mathbf{Ax}|\mathbf{Bx}) \leq \text{rank}\left(\begin{bmatrix} \mathbf{A} \\ \mathbf{B} \end{bmatrix}\right) - \text{rank}(\mathbf{B})$, *where the equality holds when the entries of* $\mathbf{x}$ *are independently and uniformly chosen from the finite field* $\mathbb{F}_q$.

Proof. For notational convenience, denote by r_A, r_B and r_{AB} the ranks of matrices $\mathbf{A}$, $\mathbf{B}$ and $\begin{bmatrix} \mathbf{A} \\ \mathbf{B} \end{bmatrix}$, respectively. Let $\mathbf{N}_1 \in \mathbb{F}_q^{l \times (l - r_{AB})}$ be a matrix whose columns form a basis for $\mathcal{N}\left(\begin{bmatrix} \mathbf{A} \\ \mathbf{B} \end{bmatrix}\right)$. Then we can find a matrix $\mathbf{N}_2$

of size $l \times (r_{AB} - r_B)$ such that the columns of $\mathbf{N}_1$ and $\mathbf{N}_2$ form a basis for $\mathcal{N}(\mathbf{B})$. Moreover, let $\mathbf{N}_3 \in \mathbb{F}_q^{l \times r_B}$ be the basis of $\mathcal{R}(\mathbf{B}^T)$. Therefore, the columns in $\begin{bmatrix} \mathbf{N}_1 & \mathbf{N}_2 & \mathbf{N}_3 \end{bmatrix}$ span the l-dimensional space. As a consequence, any given $\mathbf{x} \in \mathbb{F}_q^{l \times 1}$ can be represented as $\mathbf{x} = \begin{bmatrix} \mathbf{N}_1 & \mathbf{N}_2 & \mathbf{N}_3 \end{bmatrix} \mathbf{x}'$. Thus,

$$
\begin{aligned}
H(\mathbf{A}\mathbf{x}|\mathbf{B}\mathbf{x}) &= H\left(\mathbf{A}[\mathbf{N}_1 \ \mathbf{N}_2 \ \mathbf{N}_3]\mathbf{x}' \mid \mathbf{B}[\mathbf{N}_1 \ \mathbf{N}_2 \ \mathbf{N}_3]\mathbf{x}'\right) \\
&= H\left(\mathbf{A}\mathbf{N}_2\mathbf{x}_2' + \mathbf{A}\mathbf{N}_3\mathbf{x}_3' \mid \mathbf{B}\mathbf{N}_3\mathbf{x}_3'\right) \\
&\overset{(a)}{=} H\left(\mathbf{A}\mathbf{N}_2\mathbf{x}_2' \mid \mathbf{B}\mathbf{N}_3\mathbf{x}_3', \mathbf{x}_3'\right) \\
&\leq H(\mathbf{A}\mathbf{N}_2\mathbf{x}_2') \leq \operatorname{rank}(\mathbf{A}\mathbf{N}_2) \\
&= \operatorname{rank}(\mathbf{N}_2) = r_{AB} - r_B
\end{aligned}
$$

where (a) is true since $\mathbf{x}_3'$ can be uniquely determined from $\mathbf{B}\mathbf{N}_3\mathbf{x}_3'$. When all the entries of $\mathbf{x}$ are independently and uniformly distributed in $\mathbb{F}_q$, so are those in $\mathbf{x}'$. In this case, all the inequalities above become equalities.   □

With Lemma 1, the region given in Equations (3.3–3.9) can be equivalently written as[2]

$$R_1 \leq \operatorname{rank}(\mathbf{H}_{11}) \tag{3.10}$$

$$R_2 \leq \operatorname{rank}(\mathbf{H}_{22}) \tag{3.11}$$

$$R_1 + R_2 \leq \operatorname{rank}\left([\mathbf{H}_{11} \ \mathbf{H}_{12}]\right) + \operatorname{rank}\left(\begin{bmatrix} \mathbf{H}_{12} \\ \mathbf{H}_{22} \end{bmatrix}\right) - \operatorname{rank}(\mathbf{H}_{12}) \tag{3.12}$$

$$R_1 + R_2 \leq \operatorname{rank}\left([\mathbf{H}_{21} \ \mathbf{H}_{22}]\right) + \operatorname{rank}\left(\begin{bmatrix} \mathbf{H}_{11} \\ \mathbf{H}_{21} \end{bmatrix}\right) - \operatorname{rank}(\mathbf{H}_{21}) \tag{3.13}$$

$$R_1 + R_2 \leq \operatorname{rank}\left(\begin{bmatrix} \mathbf{H}_{11} & \mathbf{H}_{12} \\ \mathbf{H}_{21} & \mathbf{0} \end{bmatrix}\right) + \operatorname{rank}\left(\begin{bmatrix} \mathbf{H}_{21} & \mathbf{H}_{22} \\ \mathbf{0} & \mathbf{H}_{12} \end{bmatrix}\right) - \operatorname{rank}(\mathbf{H}_{21}) - \operatorname{rank}(\mathbf{H}_{12}) \tag{3.14}$$

$$
\begin{aligned}
2R_1 + R_2 \leq {} &\operatorname{rank}\left(\begin{bmatrix} \mathbf{H}_{11} \\ \mathbf{H}_{21} \end{bmatrix}\right) + \operatorname{rank}\left(\begin{bmatrix} \mathbf{H}_{21} & \mathbf{H}_{22} \\ \mathbf{0} & \mathbf{H}_{12} \end{bmatrix}\right) + \operatorname{rank}\left([\mathbf{H}_{11} \ \mathbf{H}_{12}]\right) \\
&- \operatorname{rank}(\mathbf{H}_{21}) - \operatorname{rank}(\mathbf{H}_{12})
\end{aligned} \tag{3.15}
$$

$$
\begin{aligned}
R_1 + 2R_2 \leq {} &\operatorname{rank}\left(\begin{bmatrix} \mathbf{H}_{12} \\ \mathbf{H}_{22} \end{bmatrix}\right) + \operatorname{rank}\left(\begin{bmatrix} \mathbf{H}_{11} & \mathbf{H}_{12} \\ \mathbf{H}_{21} & \mathbf{0} \end{bmatrix}\right) + \operatorname{rank}\left([\mathbf{H}_{21} \ \mathbf{H}_{22}]\right) \\
&- \operatorname{rank}(\mathbf{H}_{21}) - \operatorname{rank}(\mathbf{H}_{12}).
\end{aligned} \tag{3.16}
$$

---

[2] As the following analysis is based on the finite field of order q, the data rate in the following context is given in terms of symbols, which is equivalent to the conventional rate expressed in bits normalized by $\log_2(q)$.

Furthermore, with RLNC employed at each intermediate node over $\mathbb{F}_q$, with probability 1 as $q \to \infty$, we have [18]:

$$\text{rank}(\mathbf{H}_{11}) = k_{1-1}; \quad \text{rank}(\mathbf{H}_{12}) = k_{2-1};$$
$$\text{rank}(\mathbf{H}_{21}) = k_{1-2}; \quad \text{rank}(\mathbf{H}_{22}) = k_{2-2};$$
$$\text{rank}\left(\begin{bmatrix} \mathbf{H}_{11} \\ \mathbf{H}_{21} \end{bmatrix}\right) = k_{1-12}; \quad \text{rank}\left(\begin{bmatrix} \mathbf{H}_{12} \\ \mathbf{H}_{22} \end{bmatrix}\right) = k_{2-12}; \quad (3.17)$$
$$\text{rank}\left(\begin{bmatrix} \mathbf{H}_{11} & \mathbf{H}_{12} \end{bmatrix}\right) = k_{12-1}; \quad \text{rank}\left(\begin{bmatrix} \mathbf{H}_{21} & \mathbf{H}_{22} \end{bmatrix}\right) = k_{12-2}.$$

The remaining matrix ranks on the right hand sides of Equations (3.10–3.16) cannot be explicitly represented in terms of the min-cuts of the network. However, they can be reformulated into equivalent forms involving the min-cuts by introducing a decomposition of the interfering transition matrices $\mathbf{H}_{12}$ and $\mathbf{H}_{21}$. Taking the transition matrix $\mathbf{H}_{12}$ with size $k_{12-1} \times k_{2-12}$ and rank k_{2-1} as an example, we let

- $\mathbf{V}_{10} \in \mathbb{F}_q^{k_{2-12} \times (k_{2-12}-k_{2-1})}$ be a matrix whose columns span the null space of $\mathbf{H}_{12}$, denoted as $\mathcal{N}(\mathbf{H}_{12})$,
- $\mathbf{V}_{11} \in \mathbb{F}_q^{k_{2-12} \times k_{2-1}}$ be a matrix such that $\mathcal{R}(\mathbf{V}_{11})$ and $\mathcal{R}(\mathbf{V}_{10})$ are complementary subspaces ([26], p. 90).
- $\mathbf{U}_{10} \in \mathbb{F}_q^{k_{12-1} \times (k_{12-1}-k_{2-1})}$ be a matrix whose columns span the left null space of $\mathbf{H}_{12}$, denoted as $\mathcal{N}(\mathbf{H}_{12}^T)$, and
- $\mathbf{U}_{11} \in \mathbb{F}_q^{k_{12-1} \times k_{2-1}}$ be a matrix such that $\mathcal{R}(\mathbf{U}_{11})$ and $\mathcal{R}(\mathbf{U}_{10})$ are complementary subspaces.

Let $\mathbf{U}_1 = \begin{bmatrix} \mathbf{U}_{11} & \mathbf{U}_{10} \end{bmatrix}$ and $\mathbf{V}_1 = \begin{bmatrix} \mathbf{V}_{11} & \mathbf{V}_{10} \end{bmatrix}$. Following from the definition of the complementary subspaces ([26], p. 90), we have $\dim(\mathcal{R}(\mathbf{U}_{10}) + \mathcal{R}(\mathbf{U}_{11})) = k_{12-1}$ and $\dim(\mathcal{R}(\mathbf{V}_{10}) + \mathcal{R}(\mathbf{V}_{11})) = k_{2-12}$. $\mathbf{U}_1$ and $\mathbf{V}_1$ are full rank square matrices and hence are invertible. Let $\Lambda_1 = \mathbf{U}_1^T \mathbf{H}_{12} \mathbf{V}_1$, we then have

$$\Lambda_1 = \begin{bmatrix} \mathbf{U}_{11}^T \\ \mathbf{U}_{10}^T \end{bmatrix} \mathbf{H}_{12} \begin{bmatrix} \mathbf{V}_{11} & \mathbf{V}_{10} \end{bmatrix}$$
$$= \begin{bmatrix} \mathbf{U}_{11}^T \mathbf{H}_{12} \mathbf{V}_{11} & \mathbf{U}_{11}^T \mathbf{H}_{12} \mathbf{V}_{10} \\ \mathbf{U}_{10}^T \mathbf{H}_{12} \mathbf{V}_{11} & \mathbf{U}_{10}^T \mathbf{H}_{12} \mathbf{V}_{10} \end{bmatrix}$$
$$= \begin{bmatrix} \mathbf{U}_{11}^T \mathbf{H}_{12} \mathbf{V}_{11} & \mathbf{0}_{k_{2-1} \times (k_{2-12}-k_{2-1})} \\ \mathbf{0}_{(k_{12-1}-k_{2-1}) \times k_{2-1}} & \mathbf{0}_{(k_{12-1}-k_{2-1}) \times (k_{2-12}-k_{2-1})} \end{bmatrix}.$$

Therefore, $\mathbf{H}_{12}$ can be decomposed as

$$\mathbf{H}_{12} = (\mathbf{U}_1^T)^{-1} \Lambda_1 \mathbf{V}_1^{-1}. \quad (3.18)$$

Lemma 2. *[26] Let* $\mathbf{A} \in \mathbb{F}_q^{p \times l}$ *and* $\mathbf{B} \in \mathbb{F}_q^{l \times k}$, *then we have* $\mathrm{rank}(\mathbf{AB}) = \mathrm{rank}(\mathbf{A}) - \dim(\mathcal{R}(\mathbf{A}^T) \cap \mathcal{N}(\mathbf{B}^T)) = \mathrm{rank}(\mathbf{B}) - \dim(\mathcal{N}(\mathbf{A}) \cap \mathcal{R}(\mathbf{B}))$.

Proof. Please refer to Fact 2.10.14 on page 116 in [26]. □

Lemma 3. *The* $k_{2-1} \times k_{2-1}$ *square matrix* $\mathbf{D}_{12} \triangleq \mathbf{U}_{11}^T \mathbf{H}_{12} \mathbf{V}_{11}$ *is nonsingular.*

Proof. To prove Lemma 3, we need to show that $\mathrm{rank}(\mathbf{D}_{12}) = k_{2-1}$. First, according to Lemma 2, $\mathrm{rank}(\mathbf{U}_{11}^T \mathbf{H}_{12})$ can be computed as

$$\mathrm{rank}(\mathbf{U}_{11}^T \mathbf{H}_{12}) = \mathrm{rank}(\mathbf{U}_{11}^T) - \dim(\mathcal{R}(\mathbf{U}_{11}) \cap \mathcal{N}(\mathbf{H}_{12}^T))$$
$$= \mathrm{rank}(\mathbf{U}_{11}) - \dim(\mathcal{R}(\mathbf{U}_{11}) \cap \mathcal{R}(\mathbf{U}_{10}))$$
$$\overset{(a)}{=} \mathrm{rank}(\mathbf{U}_{11}) = k_{2-1}$$

where (a) follows since $\mathbf{U}_{10}$ and $\mathbf{U}_{11}$ are complementary subspaces. Therefore, we have $\dim(\mathcal{N}(\mathbf{U}_{11}^T \mathbf{H}_{12})) = k_{2-12} - \mathrm{rank}(\mathbf{U}_{11}^T \mathbf{H}_{12}) = k_{2-12} - k_{2-1}$. According to Lemma 2.4.1 on page 94 of [26], we have $\mathcal{N}(\mathbf{H}_{12}) \subseteq \mathcal{N}(\mathbf{U}_{11}^T \mathbf{H}_{12})$. Since $\dim(\mathcal{N}(\mathbf{H}_{12})) = \dim(\mathcal{N}(\mathbf{U}_{11}^T \mathbf{H}_{12})) = k_{2-12} - k_{2-1}$, we must have $\mathcal{N}(\mathbf{H}_{12}) = \mathcal{N}(\mathbf{U}_{11}^T \mathbf{H}_{12})$. Hence, by applying Lemma 2, we have

$$\mathrm{rank}(\mathbf{D}_{12}) = \mathrm{rank}(\mathbf{U}_{11}^T \mathbf{H}_{12} \mathbf{V}_{11})$$
$$= \mathrm{rank}(\mathbf{V}_{11}) - \dim(\mathcal{N}(\mathbf{U}_{11}^T \mathbf{H}_{12}) \cap \mathcal{R}(\mathbf{V}_{11}))$$
$$= k_{2-1} - \dim(\mathcal{N}(\mathbf{H}_{12}) \cap \mathcal{R}(\mathbf{V}_{11}))$$
$$= k_{2-1} - \dim(\mathcal{R}(\mathbf{V}_{10}) \cap \mathcal{R}(\mathbf{V}_{11}))$$
$$= k_{2-1},$$

where the last equality follows since $\mathcal{R}(\mathbf{V}_{10})$ and $\mathcal{R}(\mathbf{V}_{11})$ are complementary subspaces. This thus completes the proof of Lemma 3. □

Similarly, $\mathbf{H}_{21}$ can be decomposed as

$$\mathbf{H}_{21} = (\mathbf{U}_2^T)^{-1} \Lambda_2 \mathbf{V}_2^{-1} = \begin{bmatrix} \mathbf{U}_{21}^T \\ \mathbf{U}_{20}^T \end{bmatrix}^{-1} \begin{bmatrix} \mathbf{D}_{21} & \mathbf{0} \\ \mathbf{0} & \mathbf{0} \end{bmatrix} \begin{bmatrix} \mathbf{V}_{21} & \mathbf{V}_{20} \end{bmatrix}^{-1}. \qquad (3.19)$$

Lemma 4. *With the above decompositions of* $\mathbf{H}_{12}$ *and* $\mathbf{H}_{21}$, *we have*

$$\mathrm{rank}\left(\begin{bmatrix} \mathbf{H}_{11} & \mathbf{H}_{12} \\ \mathbf{H}_{21} & \mathbf{0} \end{bmatrix} \right) = \mathrm{rank}(\mathbf{U}_{10}^T \mathbf{H}_{11} \mathbf{V}_{20}) + k_{1-2} + k_{2-1}$$

$$\mathrm{rank}\left(\begin{bmatrix} \mathbf{H}_{21} & \mathbf{H}_{22} \\ \mathbf{0} & \mathbf{H}_{12} \end{bmatrix} \right) = \mathrm{rank}(\mathbf{U}_{20}^T \mathbf{H}_{22} \mathbf{V}_{10}) + k_{1-2} + k_{2-1}.$$

Proof. Since matrices $\mathbf{U}_1, \mathbf{U}_2, \mathbf{V}_1$, and $\mathbf{V}_2$ are non-singular, the block-diagonal matrices $\begin{bmatrix} \mathbf{U}_1^T & \mathbf{0} \\ \mathbf{0} & \mathbf{U}_2^T \end{bmatrix}$ and $\begin{bmatrix} \mathbf{V}_2 & \mathbf{0} \\ \mathbf{0} & \mathbf{V}_1 \end{bmatrix}$ are non-singular as well. With the fact that the rank of a matrix remains the same after multiplying a non-singular matrix, we have

$$
\begin{aligned}
\mathrm{rank} \begin{bmatrix} \mathbf{H}_{11} & \mathbf{H}_{12} \\ \mathbf{H}_{21} & \mathbf{0} \end{bmatrix} &= \mathrm{rank} \left(\begin{bmatrix} \mathbf{U}_1^T & \mathbf{0} \\ \mathbf{0} & \mathbf{U}_2^T \end{bmatrix} \begin{bmatrix} \mathbf{H}_{11} & \mathbf{H}_{12} \\ \mathbf{H}_{21} & \mathbf{0} \end{bmatrix} \begin{bmatrix} \mathbf{V}_2 & \mathbf{0} \\ \mathbf{0} & \mathbf{V}_1 \end{bmatrix} \right) \\
&= \mathrm{rank} \begin{bmatrix} \mathbf{U}_1^T \mathbf{H}_{11} \mathbf{V}_2 & \Lambda_1 \\ \Lambda_2 & \mathbf{0} \end{bmatrix} \\
&\overset{(a)}{=} \mathrm{rank} \begin{bmatrix} \mathbf{0} & \mathbf{0} & \mathbf{D}_{12} \\ \mathbf{0} & \mathbf{U}_{10}^T \mathbf{H}_{11} \mathbf{V}_{20} & \mathbf{0} \\ \mathbf{D}_{21} & \mathbf{0} & \mathbf{0} \end{bmatrix} \\
&= \mathrm{rank}(\mathbf{U}_{10}^T \mathbf{H}_{11} \mathbf{V}_{20}) + k_{1-2} + k_{2-1},
\end{aligned}
$$

where (a) can be obtained by elementary row and column operations with the fact that $\mathbf{D}_{12}$ and $\mathbf{D}_{21}$ are non-singular square matrices. With similar techniques, we can show that $\mathrm{rank} \left(\begin{bmatrix} \mathbf{H}_{21} & \mathbf{H}_{22} \\ \mathbf{0} & \mathbf{H}_{12} \end{bmatrix} \right) = \mathrm{rank}(\mathbf{U}_{20}^T \mathbf{H}_{22} \mathbf{V}_{10}) + k_{1-2} + k_{2-1}$. $\qquad \square$

With Equation (3.17) and Lemma 4, Equations (3.10–3.16) can be equivalently expressed as

$$R_1 \leq k_{1-1} \tag{3.20}$$

$$R_2 \leq k_{2-2} \tag{3.21}$$

$$R_1 + R_2 \leq k_{12-1} + k_{2-12} - k_{2-1} \tag{3.22}$$

$$R_1 + R_2 \leq k_{12-2} + k_{1-12} - k_{1-2} \tag{3.23}$$

$$R_1 + R_2 \leq \mathrm{rank}(\mathbf{U}_{10}^T \mathbf{H}_{11} \mathbf{V}_{20}) + \mathrm{rank}(\mathbf{U}_{20}^T \mathbf{H}_{22} \mathbf{V}_{10}) + k_{1-2} + k_{2-1} \tag{3.24}$$

$$2R_1 + R_2 \leq k_{12-1} + k_{1-12} + \mathrm{rank}(\mathbf{U}_{20}^T \mathbf{H}_{22} \mathbf{V}_{10}) \tag{3.25}$$

$$R_1 + 2R_2 \leq k_{12-2} + k_{2-12} + \mathrm{rank}(\mathbf{U}_{10}^T \mathbf{H}_{11} \mathbf{V}_{20}) \tag{3.26}$$

Given a rate pair (R_1, R_2) that satisfies the inequalities in Equations (3.20–3.26), we will next show that it can be achieved by applying the linear scheme depicted in Figure 3.2. First, with $\mathbf{H}_{12}$ and $\mathbf{H}_{21}$ decomposed as in Equations (3.18) and (3.19), respectively, and by absorbing $\mathbf{V}_j^{-1}$ into the input vector $\mathbf{x}_i$ and multiplying the output vector $\mathbf{y}_i$ by $\mathbf{U}_i^T$, $i, j \in \{1, 2\}$, $i \neq j$, the channel model given in Equation (3.2) can be equivalently written as

$$\begin{aligned} \mathbf{y}_1' &= \mathbf{U}_1^T \mathbf{y}_1 = \mathbf{H}_{11}' \mathbf{x}_1' + \Lambda_1 \mathbf{x}_2' \\ \mathbf{y}_2' &= \mathbf{U}_2^T \mathbf{y}_2 = \mathbf{H}_{22}' \mathbf{x}_2' + \Lambda_2 \mathbf{x}_1', \end{aligned} \tag{3.27}$$

where $\mathbf{x}_1' = \mathbf{V}_2^{-1} \mathbf{x}_1$, $\mathbf{x}_2' = \mathbf{V}_1^{-1} \mathbf{x}_2$, $\mathbf{H}_{11}' = \mathbf{U}_1^T \mathbf{H}_{11} \mathbf{V}_2$, and $\mathbf{H}_{22}' = \mathbf{U}_2^T \mathbf{H}_{22} \mathbf{V}_1$. The advantage of the equivalent channel model Equation (3.27) is that it results in block-diagonal interfering channel matrices Λ_1 and Λ_2, which are easier to deal with. To determine the input signal vectors $\mathbf{x}_1$ and $\mathbf{x}_2$, it is then sufficient to obtain $\mathbf{x}_1'$ and $\mathbf{x}_2'$ since they are related by the invertible linear transformations $\mathbf{x}_1 = \mathbf{V}_2 \mathbf{x}_1'$ and $\mathbf{x}_2 = \mathbf{V}_1 \mathbf{x}_2'$.

Motivated by the rate-splitting technique used in the celebrated Han-Kobayashi schemes in ICs [25], we decompose the R_1 symbols in $\mathbf{d}_1$ into two parts: the common information $\mathbf{d}_{1c} \in \mathbb{F}_q^{R_{1c}}$, which is decodable at both t_1 and t_2, and the private information $\mathbf{d}_{1p} \in \mathbb{F}_q^{R_{1p}}$, which is decodable at t_1 only. We then have $\mathbf{d}_1 = \begin{bmatrix} \mathbf{d}_{1c} \\ \mathbf{d}_{1p} \end{bmatrix}$ and $R_1 = R_{1c} + R_{1p}$. To map $\mathbf{d}_1$ to the k_{1-12}-dimensional input vector $\mathbf{x}_1'$, a block-diagonal linear precoding is applied as

$$\mathbf{x}_1' = \begin{bmatrix} \mathbf{E}_{1c} & \mathbf{0} \\ \mathbf{0} & \mathbf{E}_{1p} \end{bmatrix} \mathbf{d}_1 = \begin{bmatrix} \mathbf{E}_{1c} \mathbf{d}_{1c} \\ \mathbf{E}_{1p} \mathbf{d}_{1p} \end{bmatrix}, \tag{3.28}$$

where $\mathbf{E}_{1c} \in \mathbb{F}_q^{k_{1-2} \times R_{1c}}$ and $\mathbf{E}_{1p} \in \mathbb{F}_q^{(k_{1-12} - k_{1-2}) \times R_{1p}}$ are randomly and independently generated. Note that with the above block-diagonal precoding, the common and private symbols $\mathbf{d}_{1c}$ and $\mathbf{d}_{1p}$ are constrained to the first k_{1-2} and the last $(k_{1-12} - k_{1-2})$ components of $\mathbf{x}_1'$, respectively. As such, the private symbols $\mathbf{d}_{1p}$ will not interfere with the received signal vector $\mathbf{y}_2'$ at t_2, and at the same time, the common symbols $\mathbf{d}_{1c}$ can be decoded at both t_1 and t_2 if the rate pair (R_1, R_2) satisfies the inequalities given in Equations (3.20–3.26), as will become clear later.

Similarly, $\mathbf{x}_2'$ can be obtained as $\mathbf{x}_2' = \begin{bmatrix} \mathbf{E}_{2c} \mathbf{d}_{2c} \\ \mathbf{E}_{2p} \mathbf{d}_{2p} \end{bmatrix}$, where $\mathbf{d}_{2c} \in \mathbb{F}_q^{R_{2c}}$, $\mathbf{d}_{2p} \in \mathbb{F}_q^{R_{2p}}$, $\mathbf{E}_{2c} \in \mathbb{F}_q^{k_{2-1} \times R_{2c}}$, and $\mathbf{E}_{2p} \in \mathbb{F}_q^{(k_{2-12} - k_{2-1}) \times R_{2p}}$. $\mathbf{E}_{2c}$ and $\mathbf{E}_{2p}$ are randomly and independently generated.

The output $\mathbf{y}'_1$ at t_1 given in Equation (3.27) can then be written as

$$\mathbf{y}'_1 = \mathbf{U}_1^T \mathbf{H}_{11} \mathbf{V}_2 \mathbf{x}'_1 + \mathbf{\Lambda}_1 \mathbf{x}'_2$$

$$= \begin{bmatrix} \mathbf{U}_{11}^T \mathbf{H}_{11} \mathbf{V}_{21} & \mathbf{U}_{11}^T \mathbf{H}_{11} \mathbf{V}_{20} \\ \mathbf{U}_{10}^T \mathbf{H}_{11} \mathbf{V}_{21} & \mathbf{U}_{10}^T \mathbf{H}_{11} \mathbf{V}_{20} \end{bmatrix} \begin{bmatrix} \mathbf{E}_{1c} \mathbf{d}_{1c} \\ \mathbf{E}_{1p} \mathbf{d}_{1p} \end{bmatrix} + \begin{bmatrix} \mathbf{D}_{12} & \mathbf{0} \\ \mathbf{0} & \mathbf{0} \end{bmatrix} \begin{bmatrix} \mathbf{E}_{2c} \mathbf{d}_{2c} \\ \mathbf{E}_{2p} \mathbf{d}_{2p} \end{bmatrix}$$

$$= \begin{bmatrix} \mathbf{U}_{11}^T \mathbf{H}_{11} \mathbf{V}_{21} & \mathbf{U}_{11}^T \mathbf{H}_{11} \mathbf{V}_{20} & \mathbf{D}_{12} \\ \mathbf{U}_{10}^T \mathbf{H}_{11} \mathbf{V}_{21} & \mathbf{U}_{10}^T \mathbf{H}_{11} \mathbf{V}_{20} & \mathbf{0} \end{bmatrix} \begin{bmatrix} \mathbf{E}_{1c} \mathbf{d}_{1c} \\ \mathbf{E}_{1p} \mathbf{d}_{1p} \\ \mathbf{E}_{2c} \mathbf{d}_{2c} \end{bmatrix}. \tag{3.29}$$

Equation (3.29) clearly shows that the private symbols in vector $\mathbf{d}_{2p}$ transmitted by s_2 does not interfere with $\mathbf{y}'_1$. Furthermore, the desired symbols $\mathbf{d}_{1c}$ and $\mathbf{d}_{1p}$ can be decoded at t_1 if the linear system equations given in Equation (3.29) are uniquely solvable.

By symmetry, at t_2 we have

$$\mathbf{y}'_2 = \begin{bmatrix} \mathbf{U}_{21}^T \mathbf{H}_{22} \mathbf{V}_{11} & \mathbf{U}_{21}^T \mathbf{H}_{22} \mathbf{V}_{10} & \mathbf{D}_{21} \\ \mathbf{U}_{20}^T \mathbf{H}_{22} \mathbf{V}_{11} & \mathbf{U}_{20}^T \mathbf{H}_{22} \mathbf{V}_{10} & \mathbf{0} \end{bmatrix} \begin{bmatrix} \mathbf{E}_{2c} \mathbf{d}_{2c} \\ \mathbf{E}_{2p} \mathbf{d}_{2p} \\ \mathbf{E}_{1c} \mathbf{d}_{1c} \end{bmatrix}. \tag{3.30}$$

To find a sufficient condition on R_{1p}, R_{1c}, R_{2p}, and R_{2c} such that the system of linear equations given by Equations (3.29) and (3.30) are uniquely solvable, the following results will be used.

Lemma 5. *([26], p. 100) Given the linear relationship* $\mathbf{y} = \mathbf{A}\mathbf{x}$, *where* $\mathbf{A} \in \mathbb{F}_q^{p \times l}$, $\mathbf{x} \in \mathbb{F}_q^l$, *and* $\mathbf{y} \in \mathbb{F}_q^p$, $\mathbf{x}$ *can be uniquely determined from* $\mathbf{y}$ *if and only if* $\mathbf{A}$ *is of full column rank, i.e.,* $\mathrm{rank}(\mathbf{A}) = l$.

For notational convenience, let

$$\mathbf{M}_1 \triangleq \begin{bmatrix} \mathbf{U}_{11}^T \mathbf{H}_{11} \mathbf{V}_{21} & \mathbf{U}_{11}^T \mathbf{H}_{11} \mathbf{V}_{20} & \mathbf{D}_{12} \\ \mathbf{U}_{10}^T \mathbf{H}_{11} \mathbf{V}_{21} & \mathbf{U}_{10}^T \mathbf{H}_{11} \mathbf{V}_{20} & \mathbf{0} \end{bmatrix} \begin{bmatrix} \mathbf{E}_{1c} & \mathbf{0} & \mathbf{0} \\ \mathbf{0} & \mathbf{E}_{1p} & \mathbf{0} \\ \mathbf{0} & \mathbf{0} & \mathbf{E}_{2c} \end{bmatrix}$$

$$= \begin{bmatrix} \begin{pmatrix} \mathbf{U}_{11}^T \mathbf{H}_{11} \mathbf{V}_{21} \\ \mathbf{U}_{10}^T \mathbf{H}_{11} \mathbf{V}_{21} \end{pmatrix} \mathbf{E}_{1c} & \begin{pmatrix} \mathbf{U}_{11}^T \mathbf{H}_{11} \mathbf{V}_{20} \\ \mathbf{U}_{10}^T \mathbf{H}_{11} \mathbf{V}_{20} \end{pmatrix} \mathbf{E}_{1p} & \begin{pmatrix} \mathbf{D}_{12} \\ \mathbf{0} \end{pmatrix} \mathbf{E}_{2c} \end{bmatrix}. \tag{3.31}$$

Hence, Equation (3.29) can be written as $\mathbf{y}'_1 = \mathbf{M}_1 \begin{bmatrix} \mathbf{d}_{1c} \\ \mathbf{d}_{1p} \\ \mathbf{d}_{2c} \end{bmatrix}$.

According to Lemma 5, the sink node t_1 is able to decode $\mathbf{d}_{1c}$, $\mathbf{d}_{1p}$, and $\mathbf{d}_{2c}$ if $\mathbf{M}_1$ is of full column rank. Next, we will find a sufficient condition on the data rates R_{1c}, R_{1p}, and R_{2c} such that $\mathbf{M}_1$ has full column rank.

Lemma 6. *Let* $\mathbf{A}_1 \in \mathbb{F}_q^{p \times l_1}$, $\mathbf{A}_2 \in \mathbb{F}_q^{p \times l_2}$, *and* $\mathbf{A}_3 \in \mathbb{F}_q^{p \times l_3}$ *be given matrices, and* $\mathbf{E}_1 \in \mathbb{F}_q^{l_1 \times k_1}$, $\mathbf{E}_2 \in \mathbb{F}_q^{l_2 \times k_2}$, *and* $\mathbf{E}_3 \in \mathbb{F}_q^{l_3 \times k_3}$ *be random matrices whose entries are uniformly generated from* $\mathbb{F}_q$. *Then as* $q \to \infty$, rank$([\mathbf{A}_1\mathbf{E}_1 \quad \mathbf{A}_2\mathbf{E}_2 \quad \mathbf{A}_3\mathbf{E}_3]) = k_1 + k_2 + k_3$ *holds with high probability if the following conditions are satisfied:*

- rank$(\mathbf{A}_1) \geq k_1$
- rank$(\mathbf{A}_2) \geq k_2$
- rank$(\mathbf{A}_3) \geq k_3$
- rank$([\mathbf{A}_1 \quad \mathbf{A}_2]) \geq k_1 + k_2$
- rank$([\mathbf{A}_1 \quad \mathbf{A}_3]) \geq k_1 + k_3$
- rank$([\mathbf{A}_2 \quad \mathbf{A}_3]) \geq k_2 + k_3$
- rank$([\mathbf{A}_1 \quad \mathbf{A}_2 \quad \mathbf{A}_3]) \geq k_1 + k_2 + k_3.$

Proof. The rank of the composite matrix, i.e., rank$([\mathbf{A}_1\mathbf{E}_1 \quad \mathbf{A}_2\mathbf{E}_2 \quad \mathbf{A}_3\mathbf{E}_3])$, can be computed by recursively applying Fact 2 in Appendix. We start from the matrix with single matrix multiplication as follows

$$
\begin{aligned}
\gamma_1 &\triangleq \mathrm{rank}\left(\begin{bmatrix}\mathbf{A}_1\mathbf{E}_1 & \mathbf{A}_2 & \mathbf{A}_3\end{bmatrix}\right) \\
&= \min\left\{\mathrm{rank}([\mathbf{A}_1 \quad \mathbf{A}_2 \quad \mathbf{A}_3]), k_1 + \mathrm{rank}([\mathbf{A}_2 \quad \mathbf{A}_3])\right\}. \quad (3.32)
\end{aligned}
$$

Then, we are ready to compute the rank of matrix $\begin{bmatrix}\mathbf{A}_1\mathbf{E}_1 & \mathbf{A}_2\mathbf{E}_2 & \mathbf{A}_3\end{bmatrix}$ by applying Fact 2 in Appendix again

$$
\begin{aligned}
\gamma_2 &\triangleq \mathrm{rank}\left(\begin{bmatrix}\mathbf{A}_1\mathbf{E}_1 & \mathbf{A}_2\mathbf{E}_2 & \mathbf{A}_3\end{bmatrix}\right) \\
&= \min\left\{\gamma_1, k_2 + \mathrm{rank}([\mathbf{A}_1\mathbf{E}_1 \quad \mathbf{A}_3])\right\} \\
&= \min\left\{\gamma_1, k_2 + \mathrm{rank}([\mathbf{A}_1 \quad \mathbf{A}_3], k_1 + k_2 + \mathrm{rank}(\mathbf{A}_3))\right\} \\
&= \min\left\{\begin{array}{l}\mathrm{rank}([\mathbf{A}_1 \quad \mathbf{A}_2 \quad \mathbf{A}_3]), k_1 + \mathrm{rank}([\mathbf{A}_2 \quad \mathbf{A}_3]), \\ k_2 + \mathrm{rank}([\mathbf{A}_1 \quad \mathbf{A}_3]), k_1 + k_2 + \mathrm{rank}(\mathbf{A}_3)\end{array}\right\}. \quad (3.33)
\end{aligned}
$$

In a similar manner, we have

$$
\begin{aligned}
&\mathrm{rank}([\mathbf{A}_1\mathbf{E}_1 \quad \mathbf{A}_2\mathbf{E}_2 \quad \mathbf{A}_3\mathbf{E}_3]) \\
&= \min\left\{\gamma_2, k_3 + \mathrm{rank}([\mathbf{A}_1\mathbf{E}_1 \quad \mathbf{A}_2\mathbf{E}_2])\right\} \\
&= \min\left\{\gamma_2, k_3 + \mathrm{rank}([\mathbf{A}_1\mathbf{E}_1 \quad \mathbf{A}_2]), k_2 + k_3 + \mathrm{rank}([\mathbf{A}_1\mathbf{E}_1])\right\} \\
&= \min\left\{\begin{array}{l}\gamma_2, k_3 + \mathrm{rank}([\mathbf{A}_1 \quad \mathbf{A}_2]), k_1 + k_3 + \mathrm{rank}(\mathbf{A}_2), k_2 + k_3 + \mathrm{rank}(\mathbf{A}_1), \\ k_1 + k_2 + k_3\end{array}\right\}
\end{aligned}
$$

$$\begin{aligned}
&= \min \left\{ \begin{array}{l}
\text{rank}([\mathbf{A}_1 \quad \mathbf{A}_2 \quad \mathbf{A}_3]), k_1 + \text{rank}([\mathbf{A}_2 \quad \mathbf{A}_3]), k_2 + \text{rank}([\mathbf{A}_1 \quad \mathbf{A}_3]), \\
k_3 + \text{rank}([\mathbf{A}_1 \quad \mathbf{A}_2]), k_1 + k_2 + \text{rank}(\mathbf{A}_3), k_1 + k_3 + \text{rank}(\mathbf{A}_2), \\
k_2 + k_3 + \text{rank}(\mathbf{A}_1), k_1 + k_2 + k_3
\end{array} \right\} \\
&= k_1 + k_2 + k_3, \hspace{6cm} (3.34)
\end{aligned}$$

where the last equality follows from the conditions given in Lemma 6. □

By applying Lemma 6 to Equation (3.31), a sufficient condition for $\mathbf{M}_1$ to be of full column rank, and hence Equation (3.29) is uniquely solvable, is given by

$$R_{1c} \le \text{rank}\left(\begin{bmatrix} \mathbf{U}_{11}^T \mathbf{H}_{11} \mathbf{V}_{21} \\ \mathbf{U}_{10}^T \mathbf{H}_{11} \mathbf{V}_{21} \end{bmatrix} \right) = \text{rank}(\mathbf{U}_1^T \mathbf{H}_{11} \mathbf{V}_{21}) \stackrel{(a)}{=} \text{rank}(\mathbf{H}_{11} \mathbf{V}_{21}) \tag{3.35}$$

$$R_{1p} \le \text{rank}\left(\begin{bmatrix} \mathbf{U}_{11}^T \mathbf{H}_{11} \mathbf{V}_{20} \\ \mathbf{U}_{10}^T \mathbf{H}_{11} \mathbf{V}_{20} \end{bmatrix} \right) = \text{rank}(\mathbf{U}_1^T \mathbf{H}_{11} \mathbf{V}_{20}) \stackrel{(b)}{=} k_{1-12} - k_{1-2} \tag{3.36}$$

$$R_{2c} \le \text{rank}\left(\begin{bmatrix} \mathbf{D}_{12} \\ \mathbf{0} \end{bmatrix} \right) = k_{2-1} \tag{3.37}$$

$$R_{1c} + R_{1p} \le \text{rank}(\mathbf{U}_1^T \mathbf{H}_{11} \mathbf{V}_2) = k_{1-1} \tag{3.38}$$

$$R_{1c} + R_{2c} \le \text{rank}\left(\begin{bmatrix} \mathbf{U}_{11}^T \mathbf{H}_{11} \mathbf{V}_{21} & \mathbf{D}_{12} \\ \mathbf{U}_{10}^T \mathbf{H}_{11} \mathbf{V}_{21} & \mathbf{0} \end{bmatrix} \right) \stackrel{(c)}{=} k_{2-1} + \text{rank}(\mathbf{U}_{10}^T \mathbf{H}_{11} \mathbf{V}_{21}) \tag{3.39}$$

$$R_{1p} + R_{2c} \le \text{rank}\left(\begin{bmatrix} \mathbf{U}_{11}^T \mathbf{H}_{11} \mathbf{V}_{20} & \mathbf{D}_{12} \\ \mathbf{U}_{10}^T \mathbf{H}_{11} \mathbf{V}_{20} & \mathbf{0} \end{bmatrix} \right) \stackrel{(d)}{=} k_{2-1} + \text{rank}(\mathbf{U}_{10}^T \mathbf{H}_{11} \mathbf{V}_{20}) \tag{3.40}$$

$$R_{1c} + R_{1p} + R_{2c} \le \text{rank}\left(\begin{bmatrix} \mathbf{U}_{11}^T \mathbf{H}_{11} \mathbf{V}_{21} & \mathbf{U}_{11}^T \mathbf{H}_{11} \mathbf{V}_{20} & \mathbf{D}_{12} \\ \mathbf{U}_{10}^T \mathbf{H}_{11} \mathbf{V}_{21} & \mathbf{U}_{10}^T \mathbf{H}_{11} \mathbf{V}_{20} & \mathbf{0} \end{bmatrix} \right) \stackrel{(e)}{=} k_{12-1}, \tag{3.41}$$

where (a) holds since $\mathbf{U}_1$ is invertible, and (b) follows from Lemma 2 since

$$\begin{aligned}
\text{rank}(\mathbf{U}_1^T \mathbf{H}_{11} \mathbf{V}_{20}) &= \text{rank}(\mathbf{H}_{11} \mathbf{V}_{20}) \\
&= \text{rank}(\mathbf{V}_{20}) - \dim(\mathcal{N}(\mathbf{H}_{11}) \cap \mathcal{R}(\mathbf{V}_{20})) \\
&= \text{rank}(\mathbf{V}_{20}) - \dim(\mathcal{N}(\mathbf{H}_{11}) \cap \mathcal{N}(\mathbf{H}_{21})) \\
&\stackrel{(f)}{=} \text{rank}(\mathbf{V}_{20}) \\
&= k_{1-12} - k_{1-2}, \hspace{4cm} (3.42)
\end{aligned}$$

where (f) follows from Equation (3.17), i.e., $\begin{bmatrix} \mathbf{H}_{11} \\ \mathbf{H}_{21} \end{bmatrix}$ is of full column rank. Moreover, (c) and (d) are obtained with elementary column operations since $\mathbf{D}_{12}$ is nonsingular as given in Lemma 3; and (e) can be shown with elementary column operations together with a proof similar to that for (b), i.e.,

$$
\begin{aligned}
\text{rank} & \left(\begin{bmatrix} \mathbf{U}_{11}^T \mathbf{H}_{11} \mathbf{V}_{21} & \mathbf{U}_{11}^T \mathbf{H}_{11} \mathbf{V}_{20} & \mathbf{D}_{12} \\ \mathbf{U}_{10}^T \mathbf{H}_{11} \mathbf{V}_{21} & \mathbf{U}_{10}^T \mathbf{H}_{11} \mathbf{V}_{20} & \mathbf{0} \end{bmatrix} \right) \\
& = \text{rank}(\mathbf{U}_{10}^T \mathbf{H}_{11} \mathbf{V}_2) + \text{rank}(\mathbf{D}_{12}) \\
& = \text{rank}(\mathbf{U}_{10}^T \mathbf{H}_{11}) + k_{2-1} \\
& = \text{rank}(\mathbf{U}_{10}) - \dim(\mathcal{R}(\mathbf{U}_{10}) \cap \mathcal{N}(\mathbf{H}_{11}^T)) + k_{2-1} \\
& = k_{12-1} - \dim(\mathcal{N}(\mathbf{H}_{12}^T) \cap \mathcal{N}(\mathbf{H}_{11}^T)) \\
& = k_{12-1}
\end{aligned}
\tag{3.43}
$$

By symmetry, a sufficient condition for sink t_2 to successfully decode $\mathbf{d}_{2c}$, $\mathbf{d}_{2p}$, and $\mathbf{d}_{1c}$ is given by

$$R_{2c} \leq \text{rank}(\mathbf{H}_{22} \mathbf{V}_{11}) \tag{3.44}$$

$$R_{2p} \leq k_{2-12} - k_{2-1} \tag{3.45}$$

$$R_{1c} \leq k_{1-2} \tag{3.46}$$

$$R_{2c} + R_{2p} \leq k_{2-2} \tag{3.47}$$

$$R_{2c} + R_{1c} \leq k_{1-2} + \text{rank}(\mathbf{U}_{20}^T \mathbf{H}_{22} \mathbf{V}_{11}) \tag{3.48}$$

$$R_{2p} + R_{1c} \leq k_{1-2} + \text{rank}(\mathbf{U}_{20}^T \mathbf{H}_{22} \mathbf{V}_{10}) \tag{3.49}$$

$$R_{2c} + R_{2p} + R_{1c} \leq k_{12-2} \tag{3.50}$$

By substituting with $R_1 = R_{1c} + R_{1p}$ and $R_2 = R_{2c} + R_{2p}$, the sufficient conditions on the data rate R_1 and R_2 to ensure full decodability at the respective destinations can be obtained by applying the standard Fourier-Motzkin elimination over Equations (3.35–3.41) and Equations (3.44–3.50) and the resulting achievable rate region matches with the capacity region given in Equations (3.20–3.26). In other words, given any valid rate pair (R_1, R_2) satisfying Equations (3.20–3.26), there exists at least one valid common-private rate splitting $(R_{1c}, R_{1p}, R_{2c}, R_{2p})$ that can be found by solving $R_{1p} = R_1 - R_{1c}$ and $R_{2p} = R_2 - R_{2c}$ subject to the constraints specified in Equations (3.35–3.41) and Equations (3.44–3.50).

Remark 1. *Since Equations (3.10–3.16) is the capacity region for the linear deterministic IC given by Equation (3.2), it gives the maximum achievable* rate region for double-unicast networks with any given transition matrices $\{\mathbf{H}_{ji}\}$. In other words, for any given coding coefficients chosen by the intermediate nodes, there are no other source and sink strategies, linear or non-linear, that can outperform the proposed linear scheme described above.

Remark 2. *While the proposed linear achievability scheme has been presented mainly based on the rate region specified in Equations (3.20–3.26) under the assumption that RLNC over sufficiently large field is applied at the intermediate nodes, the same techniques can be applied to show the achievability of the rate region specified in the original form in Equations (3.10–3.16) with any given set of transition matrices resulting from any (not necessarily random) linear network coding schemes. Specifically, for general transition matrices $\mathbf{H}_{12} \in \mathbb{F}_q^{n_1 \times m_2}$ and $\mathbf{H}_{21} \in \mathbb{F}_q^{n_2 \times m_1}$ with ranks r_{12} and r_{21}, respectively, the corresponding precoding matrices in Figure 3.2 can be similarly obtained as*

- $\mathbf{V}_2 = \begin{bmatrix} \mathbf{V}_{21} & \mathbf{V}_{20} \end{bmatrix} \in \mathbb{F}_q^{m_1 \times m_1}$, *where the columns of $\mathbf{V}_{20} \in \mathbb{F}_q^{m_1 \times (m_1 - r_{21})}$ span $\mathcal{N}(\mathbf{H}_{21})$ and $\mathcal{R}(\mathbf{V}_{21})$ and $\mathcal{R}(\mathbf{V}_{20})$ are complementary subspaces.*
- $\mathbf{U}_1 = \begin{bmatrix} \mathbf{U}_{11} & \mathbf{U}_{10} \end{bmatrix} \in \mathbb{F}_q^{n_1 \times n_1}$, *where the columns of $\mathbf{U}_{10} \in \mathbb{F}_q^{n_1 \times (n_1 - r_{12})}$ span $\mathcal{N}(\mathbf{H}_{12}^T)$ and $\mathcal{R}(\mathbf{U}_{11})$ and $\mathcal{R}(\mathbf{U}_{10})$ are complementary subspaces.*
- $\mathbf{E}_{1c} \in \mathbb{F}_q^{r_{21} \times R_{1c}}$ *and* $\mathbf{E}_{1p} \in \mathbb{F}_q^{(m_1 - r_{21}) \times R_{1p}}$ *are generated randomly.*

The proposed linear achievability scheme presented above is illustrated with the following simple example.

Example 1. *Consider the network shown in Figure 3.3(a) with unit edge capacity. The min-cut values of this network are respectively given by $k_{1-1} = 2$, $k_{2-2} = 3$, $k_{2-1} = k_{1-2} = 2$, $k_{1-12} = k_{12-1} = 2$, and $k_{2-12} = k_{12-2} = 3$. Assume that random network coding coefficients have been chosen in the field of size[3] $q = 7$ with the coding coefficients at each intermediate node indicated in the figure, and the resulting transition matrices are given by:*

$$\mathbf{H}_{11} = \begin{bmatrix} 2 & 0 \\ 2 & 3 \end{bmatrix}, \mathbf{H}_{12} = \begin{bmatrix} 2 & 1 & 0 \\ 2 & 1 & 1 \end{bmatrix}, \mathbf{H}_{21} = \begin{bmatrix} 1 & 0 \\ 2 & 3 \\ 2 & 3 \end{bmatrix}, \mathbf{H}_{22} = \begin{bmatrix} 1 & 0 & 0 \\ 2 & 1 & 0 \\ 2 & 1 & 1 \end{bmatrix}.$$
$$(3.51)$$

---

[3]One commonly used field size for RLNC is 2^8. Here, a small field size is used for illustration purposes.

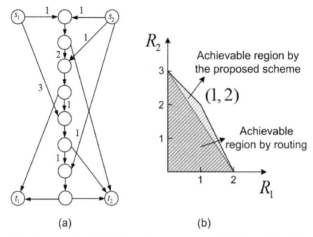

Figure 3.3   (a) An example of double-unicast networks. (b) The achievable rate regions.

Then the achievable region (after removing redundant inequalities) specified in Equations (3.10–3.16) for this particular example is given by

$$R_1 \leq 2$$
$$R_1 + R_2 \leq 3$$
$$2R_1 + R_2 \leq 4,$$

which is plotted in Figure 3.3(b). As a comparison, the achievable rate region with routing is also plotted, which is shown to be strictly smaller than that achieved with network coding.

By applying the above proposed scheme, we next give the specific designs that achieve the outmost rate pair $(R_1, R_2) = (1, 2)$. *Let* $\mathbf{d}_1 = d_{11}$ *and* $\mathbf{d}_2 = \begin{bmatrix} d_{21} & d_{22} \end{bmatrix}^T$ *be the information-bearing symbols. Then, following Equations (3.35–3.41) and Equations (3.44–3.50), the private-common rate splitting is given by* $\mathbf{d}_{1c} = d_{11}, \mathbf{d}_{1p} = \emptyset, \mathbf{d}_{2c} = d_{21}$ *and* $\mathbf{d}_{2p} = d_{22},$ *where* $\emptyset$ *denotes an empty vector/matrix.*

Based on $\mathbf{H}_{12}$ *and* $\mathbf{H}_{21}$ *given in Equation (3.51), the following component matrices can be obtained:*

$$\mathbf{U}_{11} = \begin{bmatrix} 1 & 0 \\ 1 & 1 \end{bmatrix}, \mathbf{U}_{10} = \emptyset, \mathbf{V}_{11} = \begin{bmatrix} 2 & 2 \\ 1 & 1 \\ 0 & 1 \end{bmatrix}, \mathbf{V}_{10} = \begin{bmatrix} 1 \\ 5 \\ 0 \end{bmatrix},$$

$$\mathbf{U}_{21} = \begin{bmatrix} 1 & 0 \\ 2 & 3 \\ 2 & 3 \end{bmatrix}, \mathbf{U}_{20} = \begin{bmatrix} 0 \\ 3 \\ 4 \end{bmatrix}, \mathbf{V}_{21} = \begin{bmatrix} 1 & 2 \\ 0 & 3 \end{bmatrix}, \mathbf{V}_{20} = \emptyset.$$

Assume that the following matrices are randomly generated, which will be applied as the linear precoding matrices to map the data symbols to the transmit vectors:

$$\mathbf{E}_{1c} = \begin{bmatrix} 4 \\ 3 \end{bmatrix}, \mathbf{E}_{1p} = \emptyset, \mathbf{E}_{2c} = \begin{bmatrix} 2 \\ 3 \end{bmatrix}, \mathbf{E}_{2p} = \begin{bmatrix} 3 \end{bmatrix}.$$

The transmit vector by s_1 can then be obtained as

$$\mathbf{x}_1 = \mathbf{V}_2 \mathbf{x}_1' = \mathbf{V}_{21} \mathbf{E}_{1c} \mathbf{d}_{1c} = \begin{bmatrix} 1 & 2 \\ 0 & 3 \end{bmatrix} \begin{bmatrix} 4 \\ 3 \end{bmatrix} d_{11} = \begin{bmatrix} 3d_{11} \\ 2d_{11} \end{bmatrix}$$

Similarly, we have

$$\mathbf{x}_2 = \mathbf{V}_1 \mathbf{x}_2' = \begin{bmatrix} \mathbf{V}_{11} & \mathbf{V}_{10} \end{bmatrix} \begin{bmatrix} \mathbf{E}_{2c} \mathbf{d}_{2c} \\ \mathbf{E}_{2p} \mathbf{d}_{2p} \end{bmatrix} = \begin{bmatrix} 2 & 2 & 1 \\ 1 & 1 & 5 \\ 0 & 1 & 0 \end{bmatrix} \begin{bmatrix} 2d_{21} \\ 3d_{21} \\ 3d_{22} \end{bmatrix} = \begin{bmatrix} 3d_{21} + 3d_{22} \\ 5d_{21} + d_{22} \\ 3d_{21} \end{bmatrix}$$

Therefore, the received symbol vectors at t_1 and t_2 are respectively given by

$$\mathbf{y}_1 = \mathbf{H}_{11} \begin{bmatrix} 3d_{11} \\ 2d_{11} \end{bmatrix} + \mathbf{H}_{12} \begin{bmatrix} 3d_{21} + 3d_{22} \\ 5d_{21} + d_{22} \\ 3d_{21} \end{bmatrix} = \begin{bmatrix} 6d_{11} + 4d_{21} \\ 5d_{11} \end{bmatrix} \tag{3.52}$$

$$\mathbf{y}_2 = \mathbf{H}_{22} \begin{bmatrix} 3d_{21} + 3d_{22} \\ 5d_{21} + d_{22} \\ 3d_{21} \end{bmatrix} + \mathbf{H}_{21} \begin{bmatrix} 3d_{11} \\ 2d_{11} \end{bmatrix} = \begin{bmatrix} 3d_{21} + 3d_{22} + 3d_{11} \\ 4d_{21} + 5d_{11} \\ 5d_{11} \end{bmatrix}. \tag{3.53}$$

It can be verified that both t_1 and t_2 can recover their desired symbols by solving the system of linear equations given by Equations (3.52) and (3.53), respectively.

3.3.3 An Achievable Region in Terms of Min-cuts

The achievable rate region for double-unicast networks specified in Equations (3.20–3.26) is given in terms of the min-cuts of the network, as well as the ranks of matrices $\mathbf{U}_{10}^T \mathbf{H}_{11} \mathbf{V}_{20}$ and $\mathbf{U}_{20}^T \mathbf{H}_{22} \mathbf{V}_{10}$. In this section, we derive another rate region for Equation (3.2) only in terms of the min-cuts of the network, which is sub-optimal but can be more easily computed. The following result will be used:

Lemma 7. *If the coding coefficients at the intermediate nodes are randomly selected from a sufficiently large field, we have the following inequalities:*

$$\text{rank} \left(\mathbf{U}_{10}^T \mathbf{H}_{11} \mathbf{V}_{20} \right) \geq (k_{12-1} + k_{1-12} - k_{1-1} - k_{1-2} - k_{2-1})^+, \tag{3.54}$$

$$\text{rank} \left(\mathbf{U}_{20}^T \mathbf{H}_{22} \mathbf{V}_{10} \right) \geq (k_{12-2} + k_{2-12} - k_{2-2} - k_{2-1} - k_{1-2})^+. \tag{3.55}$$

Proof. With Frobenius inequality, we have

$$\text{rank}(\mathbf{U}_{10}^T\mathbf{H}_{11}\mathbf{V}_{20}) \geq \left(\text{rank}\left(\mathbf{U}_{10}^T\mathbf{H}_{11}\right) + \text{rank}\left(\mathbf{H}_{11}\mathbf{V}_{20}\right) - \text{rank}\left(\mathbf{H}_{11}\right)\right)^+$$

$$\overset{(a)}{=} \left(k_{12-1} - k_{2-1} + k_{1-12} - k_{1-2} - k_{1-1}\right)^+,$$

where (a) follows from Lemma 2. Equation (3.55) follows by symmetry. □

By substituting Equations (3.54) and (3.55) into Equations (3.20–3.26), the following achievable rate region only in terms of min-cuts of the network can be obtained:

$$R_1 \leq k_{1-1} \tag{3.56}$$

$$R_2 \leq k_{2-2} \tag{3.57}$$

$$R_1 + R_2 \leq k_{12-1} + k_{2-12} - k_{2-1} \tag{3.58}$$

$$R_1 + R_2 \leq k_{12-2} + k_{1-12} - k_{1-2} \tag{3.59}$$

$$R_1 + R_2 \leq \left(k_{12-2} + k_{2-12} - k_{2-2} - k_{1-2} - k_{2-1}\right)^+ + k_{1-2} + k_{2-1}$$
$$+ \left(k_{12-1} + k_{1-12} - k_{1-1} - k_{1-2} - k_{2-1}\right)^+ \tag{3.60}$$

$$2R_1 + R_2 \leq \left(k_{12-2} + k_{2-12} - k_{2-2} - k_{1-2} - k_{2-1}\right)^+ + k_{12-1} + k_{1-12} \tag{3.61}$$

$$R_1 + 2R_2 \leq \left(k_{12-1} + k_{1-12} - k_{1-1} - k_{1-2} - k_{2-1}\right)^+ + k_{12-2} + k_{2-12}. \tag{3.62}$$

Example 2. *While the achievable region Equations (3.56–3.62) can be easily computed based on the min-cuts of the network, it is usually smaller than the original region specified in Equations (3.10–3.16). This is illustrated by the grail network shown in Figure 3.4.*

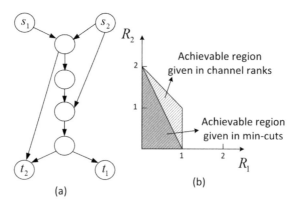

(a)   (b)

Figure 3.4   (a) The grail network. (b) The achievable regions.

In [16], the rate region specified in Equations (3.56–3.62) was shown to be strictly larger than those achievable regions obtained in [14, 18, 27].

3.3.4 Joint Routing and RLNC

Although the proposed precoding and decoding schemes in Figure 3.2 are optimal for the resulting linear deterministic IC given any set of RLNC applied at the intermediate nodes [24], there exists a gap between the rate region given in Equations (3.10–3.16) and the capacity region of double-unicast networks. This is because RLNC is in general a suboptimal strategy for the intermediate nodes. Intuitively, with RLNC, the information is mixed at the intermediate nodes to the largest extend, which results in severe interference received at the sink node. On the other hand, with the conventional routing strategy, the paths from the source to the corresponding receiver are pre-determined and the information is sent over these edge-disjoint paths. Hence, there is no interference across sessions.

In this subsection, we discuss a joint routing and RLNC (J-R-RLNC) scheme, which may achieve higher rate region compared to the schemes with either pure routing or pure RLNC. With the proposed J-R-RLNC, part of the information is routed to the receivers without network coding, to reduce the inter-session interference, and the rest is transmitted over the network via RLNC. The routing paths and the corresponding routing rates are optimized using Fourier–Motzkin elimination to maximize the overall achievable region.

Consider a general double-unicast network $G = (V, E)$, where the capacity of edge $e \in E$ is denoted by C_e. With the proposed precoding scheme and RLNC performed at the intermediate nodes, the region specified in Equations (3.56–3.62) can be achieved. To further enlarge this region by incorporating with routing, we first need to choose the proper routing paths and determine the associated routing rates. Assume that there are m paths from s_1 to t_1, denoted as $P_1^1, \ldots, P_m^1$, and n paths from s_2 to t_2, denoted as $P_1^2, \ldots, P_n^2$. Let the routing rate assigned to path P_i^1 and P_j^2 be α_i^1 and α_j^2, respectively. We then have the following constraints:

$$\alpha_i^1 \geq 0, i = 1, \ldots, m \tag{3.63}$$

$$\alpha_j^2 \geq 0, j = 1, \ldots, n \tag{3.64}$$

$$\sum_{\{i:e\in P_i^1\}} \alpha_i^1 + \sum_{\{j:e\in P_j^2\}} \alpha_j^2 \leq C_e, \forall e \in E, \tag{3.65}$$

where Equations (3.63) and (3.64) are non-negativity constraints of routing rates, and Equation (3.65) ensures that the total information routed along an edge should not exceed its capacity.

After subtracting the resource used for routing, we obtain a new network G' with the capacity for link e reduced to:

$$C'_e = C_e - \sum_{\{i:e\in P_i^1\}} \alpha_i^1 - \sum_{\{j:e\in P_j^2\}} \alpha_j^2, \forall e \in E. \qquad (3.66)$$

Denote the min-cut between the set of source s_I and the set of receivers t_J in G' by k'_{I-J}, where $I, J \subseteq \{1, 2\}$. We can obtain an achievable rate region on G' with RLNC by substituting the min-cut values into Equations (3.56–3.62), denoted as $\mathbb{R}'(R_1, R_2)$. Adding the rates achieved by routing for each session, the new achievable rate region is thus given by $\mathbb{R}'(R_1 + \sum_{i=1}^m \alpha_i^1, R_2 + \sum_{j=1}^n \alpha_j^2)$. Therefore, the routing rates $\{\alpha_i^1, \alpha_j^2\}$ can be solved by maximizing the new achievable region subject to the constraints given in Equations (3.63–3.65). This linear optimization problem can be solved by using the standard Fourier–Motzkin elimination techniques.

Example 3. *Consider the double-unicast network in Figure 3.5, where each edge has unit capacity. With RLNC performed at the intermediate nodes, the largest achievable region is given by*

$$R_1 \leq 2$$
$$R_1 + R_2 \leq 3. \qquad (3.67)$$

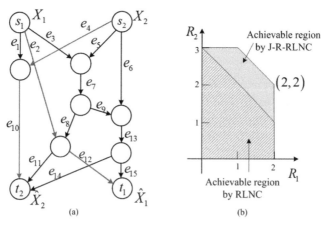

Figure 3.5   (a) A double-unicast network. (b) The achievable regions with RLNC and J-R-RLNC.

With the proposed J-R-RLNC, the achievable region can be enlarged to

$$R_1 \leq 2$$
$$R_1 \leq 3 \qquad\qquad (3.68)$$
$$R_1 + R_2 \leq 4.$$

For instance, the rate pair $(R_1, R_2) = (2, 2)$ is achieved by routing X_1 through the path $P_1^1 = \{e_2, e_{12}\}$, routing X_2 through $P_1^2 = \{e_4, e_{10}\}$ each at unit rate, and then applying RLNC over the rest of the network.

The proposed J-R-RLNC scheme can be applied to a general multi-session network where an achievable region with RLNC can be obtained. An example of applying J-R-RLNC on 2-message degraded multicast networks was given in [28]. By routing part of those information to their designated receivers, the inter-session interference can be reduced and thus the achievable region may be enlarged as compared to the one achieved by RLNC. The routing path and the corresponding rate can be determined based on the optimization result using simple Fourier–Motzkin elimination. However, the complexity of the proposed optimization problem for determining the routing rates grows exponentially with the network size. One possible future work direction is to formulate the problem with edge variables, instead of path variables, so that the computational complexity can be reduced.

3.4 Asymptotic Capacity-achieving for Multi-source Erasure Networks

Multi-source single-sink network models the communication scenario where multiple sources wish to send independent information to a common sink node. It has attracted significant interests in network computing where all the information of disjoint sources are reproduced at the receiver [29] and hence arbitrary functions can be computed. Multi-source single-sink network also provides a suitable model for the wireless sensor networks with a large number of nodes sending their respective data measurements to the fusion center for joint processing. For deterministic multi-source single-sink networks where all the links are error-free, the capacity is given by the celebrated max-flow min-cut theorem, which is achievable by simple routing [30]. However, when the links are subject to packet errors/erasures, network coding is in general needed to achieve the network capacity [31].

In practical networks, the communication links suffer from packet drops due to various reasons, such as congestion and buffer overflow. Particularly, wireless channels also experience link outage due to channel fading. Packet drops and link outage are usually modeled by packet erasures with certain erasure probabilities, and the resultant network is referred to as "erasure networks". The simplest model of erasure networks was introduced in [32], where each link corresponds to an independent memoryless erasure channel. Such a model is suitable for general wireline networks and wireless networks where orthogonal channels are used between different pairs of nodes. In [7], the broadcast nature of wireless communication has been incorporated by requiring each node to send the same information over all outgoing edges. Due to the broadcast constraint, the capacity derived in [7] provides a lower bound for the capacity of the erasure networks with orthogonal channels, where the latter is given by the min-cut of the network subject to erasures [32]. By re-defining the capacity of the edge cut, the capacity region obtained in [7] was shown to also have a nice max-flow min-cut interpretation. Another important issue for wireless transmission is the interference at the receiving nodes, i.e., the signals received at a common node from different links interfere with each other. Erasure networks with interference was considered in [33] for unicast networks and the duality between broadcast and interference has also been discussed in [33]. Later, the result in [33] was extended to multi-source single-sink erasure networks in [34]. The non-erasable networks with both broadcast at the transmitters and interference at the receivers have been studied in [35]. The most general model for erasure networks was introduced and studied in [36].

In this section, we study the multi-source single-sink erasure networks from an algebraic perspective. Specifically, we apply RLNC at the intermediate nodes over a sufficiently large number of time extensions, and model the resultant network as a linear finite-field multiple-access channel (MAC). It is shown that the capacity of the time-extended multi-source erasure network matches the capacity of this special class of MAC, which can be achieved by linear precoding at the source nodes. Hence, we provide an alternative capacity-achieving scheme for multi-source erasure networks which is much simpler than those proposed in [7, 32]. Since only linear processing are needed at all the source and intermediate nodes, the desired symbols can be decoded at the sink node by simply solving a set of linear equations.

3.4.1 The Capacity Region

A multi-source single-sink erasure network is modeled by a directed acyclic graph $G(V, E)$, as shown in Figure 3.6, where a set of source nodes $S =$

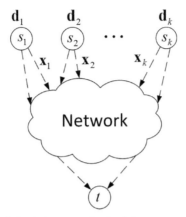

Figure 3.6 A k-source single-sink erasure network.

$\{s_1, ..., s_k\}$ want to send independent symbols $\mathbf{d}_1, ..., \mathbf{d}_k$, chosen from the finite field $\mathbb{F}_q$, to a common sink node t. We assume that each edge $e = (u, v) \in E$ is a discrete memoryless channel with unit capacity subject to an erasure probability ϵ_e, i.e., it is able to convey one symbol per time slot with probability $1 - \epsilon_e$.

It has been shown in [32] that for a discrete memoryless erasure channel with erasure probability ϵ, the effective capacity is equal to that of the corresponding deterministic channel scaled by $1 - \epsilon$. As a result, the capacity of erasure networks can be obtained based on the corresponding deterministic networks with the celebrated max-flow min-cut theorem, which leads to the following result:

Theorem 1. *The capacity region of the multi-source single-sink erasure network $G(V, E)$ as shown in Figure 3.6 is given by*

$$\mathbb{R}(G) = \left\{ \begin{array}{c} (R_i, s_i \in S) \,|\, 0 \leq \sum_{i \in I} R_i \leq \min_{V' \subseteq (V \setminus t): s_I \subseteq V'} C(V'), \\ \forall I \subseteq \{1, 2, ..., k\} \end{array} \right\}, \quad (3.69)$$

where $s_I \triangleq \{s_i, i \in I\}$ denotes the collection of the source nodes with indices in set I, and $C(V')$ is the capacity of the edge cut that separates V' from $V \setminus V'$, which is given by

$$C(V') \triangleq \sum_{\{e=(u,v): u \in V', v \in (V \setminus V')\}} (1 - \epsilon_e) \quad (3.70)$$

Proof. The proof can be easily obtained by applying the result in [4] to the multi-source single-sink erasure network; hence it is omitted here for brevity.

□

The region given in Equation (3.69) shows that the sum rate from any subset of the source nodes $s_I \subseteq S$ to t should not exceed the capacity of any edge cut that separates s_I from t. The existing achievable schemes for region Equation (3.69) are all based on information-theoretical analysis, which are difficult to be efficiently implemented in practice [7, 32].

3.4.2 Asymptotical Capacity-achieving with RLNC

In this subsection, we show that the capacity region of the multi-source single-sink erasure network, as given in Equation (3.69), is asymptotically achievable with RLNC over a sufficiently large number of time extensions. Furthermore, the sink node can decode the desired information by solving a set of linear equations in finite field.

The main idea is to model the resulting time-extended network with RLNC as a finite-field linear deterministic MAC, for which linear precoding at the source nodes is shown to be capacity-achieving.

3.4.2.1 Time-extended networks

We first show that the rate region $\mathbb{R}(G)$ given in Equation (3.69) scaled by M times, i.e., $\mathbb{R}^M(G) \triangleq M \times \mathbb{R}(G)$, is achievable for the time-extended erasure network across M time slots. Consider an arbitrary intermediate node $v \in V$ for the network $G(V, E)$ shown in Figure 3.6. Denote by $\mathbf{Z}$ the total number of packets received by v over M time slots from an edge $e \in \text{In}(v)$ with erasure probability ϵ_e. It follows that $\mathbf{Z}$ is a random number with binomial distribution, i.e., $\mathbf{Z} \sim \mathcal{B}(M, 1 - \epsilon_e)$.

Assume that there are m_i outgoing edges from source $s_i \in S$ and n incoming edges to the sink node t, i.e., $|\text{Out}(s_i)| = m_i, i = 1, ..., k$ and $|\text{In}(t)| = n$. With RLNC applied at all the intermediate nodes over M time slots, the input-output relation between the source node s_i and the sink node t can be represented by a transition matrix $\mathbf{H}_i^M$ of size $(Mn \times Mm_i)$. Based on Theorem 3 in [8], when the field size q is sufficiently large, the rank of the transition matrix equals the min-cut of the network with high probability. Furthermore, by the law of large numbers, as $M \to \infty, \mathbf{Z} \to M(1 - \epsilon_e)$. In other words, the corresponding link of edge e in the time-extended network approaches to a *deterministic* channel with capacity $M(1 - \epsilon_e)$. Therefore, the min-cut value of the erasure network can be readily determined based on its deterministic equivalent. By combining these results, we have

$$\text{rank}(\mathbf{H}_i^M) = M \left[\min_{V' \subseteq (V \backslash t) : s_i \in V'} C(V') \right], \quad \text{as } q, M \to \infty, \quad (3.71)$$

where $C(V')$ is given in Equation (3.70).

For any source subset $I \subseteq \{1, 2, ...k\}$, denote by $\mathbf{H}_I^M$ the row-concatenated matrix consisting of $\mathbf{H}_i^M, i \in I$. For instance, $\mathbf{H}_{\{1,2\}}^M = \left[\mathbf{H}_1^M \; \mathbf{H}_2^M \right]$. $\mathbf{H}_I^M$ can then be interpreted as the transition matrix from an imaginary source node s_I' to sink node t, where s_I' is connected to the set of source nodes $s_I \triangleq \{s_i, i \in I\}$ with edges of infinite capacity. Following an argument similar to that for Equation (3.71), $\forall I \subseteq \{1, 2, ..., k\}$, we have

$$\text{rank}(\mathbf{H}_I^M) = M \left[\min_{V' \subseteq (V \backslash t) : s_I \subseteq V'} C(V') \right], \quad \text{as } q, M \to \infty. \quad (3.72)$$

Therefore, to prove that the capacity region in Equation (3.69) is achievable for the original multi-source single-sink erasure network, it is tantamount to showing that the following rate region is achievable in the time-extended network

$$\mathbb{R}^M(G) = \left\{ (R_i, s_i \in S) \mid 0 \le \sum_{i \in I} R_i \le \text{rank}\left(\mathbf{H}_I^M \right), \forall I \subseteq \{1, 2, ..., k\} \right\}. \quad (3.73)$$

3.4.2.2 Linear finite-field MAC

With RLNC performed at the intermediate nodes across M time slots, the input-output relation between all the source nodes S and the sink node t can be expressed as

$$\mathbf{y}^M = \sum_{i=1}^k \mathbf{H}_i^M \mathbf{x}_i^M, \quad (3.74)$$

where $\mathbf{x}_i^M \in \mathbb{F}_q^{Mm_i}$ consists of symbols conveyed by the m_i outgoing edges of s_i over M time slots, and $\mathbf{y}^M \in \mathbb{F}_q^{Mn}$ contains the symbols received by the sink node t.

Note that the mathematical relation given in Equation (3.74) belongs to a class of MAC, which has been extensively studied in wireless communications [38]. However, in contrast to wireless channels where channel matrices are usually assumed to be generic and independent of each other, the transition matrices in Equation (3.74) are *correlated* due to the network topology constraint. Thus, the existing results for the wireless MAC cannot be directly applied to show the achievability of the rate region Equation (3.73).

In our proposed scheme to achieve the rate region Equation (3.73), the source outputs $\mathbf{x}_i^M$ in Equation (3.74) are obtained from the information-bearing symbols $\mathbf{d}_i^M$ based on random linear precoding, i.e., $\mathbf{x}_i^M = \mathbf{E}_i \mathbf{d}_i^M$, where $\mathbf{E}_i$ is the randomly generated matrix from $\mathbb{F}_q^{Mm_i \times |\mathbf{d}_i^M|}$ at source s_i, with $|\mathbf{d}_i^M|$ denoting the total number of symbols transmitted by source s_i over M time slots. As a result, the sink node t receives a linear combination of the information-bearing symbols as

$$\mathbf{y}^M = \sum_{i=1}^{k} \mathbf{H}_i^M \mathbf{E}_i \mathbf{d}_i^M, \tag{3.75}$$

which is illustrated with Figure 3.7.

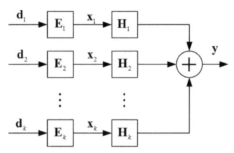

Figure 3.7   The input–output relation for a multi-source single-sink network with RLNC at the intermediate nodes and random linear precoding at the source nodes.

Define $\mathbf{G}_i = \mathbf{H}_i^M \mathbf{E}_i$. Further denote by $\mathbf{G}_I$ the row-concatenated matrix consisting of $\{\mathbf{G}_i, i \in I\}$. Clearly, t can decode all the desired symbols $\{\mathbf{d}_i^M, i = 1, ..., k\}$ if the transition matrix $\mathbf{G}_{\{1,...,k\}}$ is of full column rank, i.e., $\mathrm{rank}(\mathbf{G}_{\{1,...,k\}}) = \sum_{i=1}^{k} |\mathbf{d}_i^M|$. Our objective is to show that for any rate-tuple $\{|\mathbf{d}_1^M|, \cdots, |\mathbf{d}_k^M|\}$ within the region specified in Equation (3.73), $\mathrm{rank}(\mathbf{G}_{\{1,...,k\}}) = \sum_{i=1}^{k} |\mathbf{d}_i^M|$ is always satisfied, thus, all the desired symbols $\mathbf{d}_1^M, ..., \mathbf{d}_k^M$ can be decoded by the sink node t.

Since $\mathbf{E}_i$ in Equation (3.75) is randomly generated, from Fact 1 given in the Appendix, we have

$$\mathrm{rank}(\mathbf{G}_i) = \mathrm{rank}(\mathbf{E}_i^T (\mathbf{H}_i^M)^T) = \min\{|\mathbf{d}_i^M|, \mathrm{rank}(\mathbf{H}_i^M)\}, \tag{3.76}$$

If $\{|\mathbf{d}_1^M|, \cdots, |\mathbf{d}_k^M|\}$ is within the capacity region given in Equation (3.73), we have $|\mathbf{d}_i^M| \leq \mathrm{rank}(\mathbf{H}_i^M)$. Hence, $\mathrm{rank}(\mathbf{G}_i) = |\mathbf{d}_i^M|$, which implies that the symbols sent by each source can be decoded at the sink node t if there is no interference caused by the other sources, which is consistent with the results in [7] and [39] for unicast networks.

Lemma 8. *If* $\{|\mathbf{d}_1^M|, \cdots, |\mathbf{d}_k^M|\}$ *is within the capacity region given in Equation (3.73), we have* $\mathrm{rank}(\mathbf{G}_{\{1,\ldots,k\}}) = \sum_{i=1}^{k} |\mathbf{d}_i^M|$ *with high probability when the finite field size* $q \to \infty$.

Proof. Lemma 8 can be proved by recursively applying Fact 2 in Appendix. For convenience, define

$$\gamma_i = \mathrm{rank}\left(\begin{bmatrix} \mathbf{G}_{\{1,\ldots,i\}} & \mathbf{H}_{\{i+1,\ldots,k\}}^M \end{bmatrix}\right)$$
$$= \mathrm{rank}\left(\begin{bmatrix} \mathbf{H}_1^M \mathbf{E}_1 & \cdots & \mathbf{H}_i^M \mathbf{E}_i & \mathbf{H}_{i+1}^M & \cdots & \mathbf{H}_k^M \end{bmatrix}\right).$$

According to Fact 2, with high probability as $q \to \infty$, we have

$$\gamma_1 = \mathrm{rank}\left(\begin{bmatrix} \mathbf{H}_1^M \mathbf{E}_1 & \mathbf{H}_2^M & \cdots & \mathbf{H}_k^M \end{bmatrix}\right)$$
$$\overset{(a)}{=} \min\left\{\mathrm{rank}(\mathbf{H}_{\{1,..,k\}}^M), |\mathbf{d}_1^M| + \mathrm{rank}(\mathbf{H}_{\{2,...k\}}^M)\right\}, \qquad (3.77)$$

where (a) follows from Fact 2 with $\mathbf{A}_1 = \mathbf{H}_1^M$, $\mathbf{A}_2 = \mathbf{H}_{\{2,\ldots,k\}}^M$, and $\mathbf{B} = \mathbf{E}_1$. Similarly, γ_2 can be computed by recursively applying Fact 2 in the Appendix as

$$\gamma_2 = \mathrm{rank}\left(\begin{bmatrix} \mathbf{H}_1^M \mathbf{E}_1 & \mathbf{H}_2^M \mathbf{E}_2 & \mathbf{H}_3^M & \cdots & \mathbf{H}_k^M \end{bmatrix}\right)$$
$$= \min\left\{\gamma_1, |\mathbf{d}_2^M| + \mathrm{rank}\left(\begin{bmatrix} \mathbf{H}_1^M \mathbf{E}_1 & \mathbf{H}_3^M & \cdots & \mathbf{H}_k^M \end{bmatrix}\right)\right\}$$
$$= \min\left\{\gamma_1, |\mathbf{d}_2^M| + \mathrm{rank}\left(\mathbf{H}_{\{1,3,4,\ldots,k\}}^M\right), |\mathbf{d}_1^M| + |\mathbf{d}_2^M| + \mathrm{rank}\left(\mathbf{H}_{3,\ldots,k}^M\right)\right\}$$
$$\overset{(b)}{=} \min\left\{\begin{array}{l} \mathrm{rank}(\mathbf{H}_{\{1,\ldots,k\}}^M), |\mathbf{d}_1^M| + \mathrm{rank}(\mathbf{H}_{\{2,\ldots,k\}}^M), |\mathbf{d}_2^M| + \mathrm{rank}\left(\mathbf{H}_{\{1,3,4,\ldots,k\}}^M\right), \\ |\mathbf{d}_1^M| + |\mathbf{d}_2^M| + \mathrm{rank}\left(\mathbf{H}_{3,\ldots,k}^M\right) \end{array}\right\}$$
$$= \min_{J \subseteq \{1,2\}}\left\{\sum_{j \in J} |\mathbf{d}_j^M| + \mathrm{rank}\left(\mathbf{H}_{\{1,\ldots,k\}\setminus J}^M\right)\right\}, \qquad (3.78)$$

where (b) follows by substituting γ_1 obtained in Equation (3.77).

By applying Fact 2 recursively, we can obtain

$$\gamma_k = \text{rank}\left(\mathbf{G}_{\{1,\ldots,k\}}\right)$$

$$= \min_{J \subseteq \{1,\ldots,k\}} \left\{ \sum_{j \in J} |\mathbf{d}_j^M| + \text{rank}\left(\mathbf{H}_{\{1,\ldots,k\} \backslash J}^M\right) \right\}. \tag{3.79}$$

Since the rate tuple $\{\mathbf{d}_1, \cdots, \mathbf{d}_k\}$ is within the region specified in Equation (3.73), we have

$$\sum_{i \in I} |\mathbf{d}_i^M| \leq \text{rank}\left(\mathbf{H}_I^M\right), \forall I \subseteq \{1,\ldots,k\}, \tag{3.80}$$

which renders

$$\sum_{j \in J} |\mathbf{d}_j^M| + \text{rank}\left(\mathbf{H}_{\{1,\ldots,k\} \backslash J}^M\right) \geq \sum_{j \in J} |\mathbf{d}_j^M| + \sum_{i \in \{1,\ldots,k\} \backslash J} |\mathbf{d}_i^M|$$

$$= \sum_{i=1}^{k} |\mathbf{d}_i^M|. \tag{3.81}$$

Therefore, the value of γ_k returned by the $\min(\cdot)$ operator in Equation (3.79) will occur at $J = \emptyset$, i.e., $\gamma_k = \sum_{i=1}^{k} |\mathbf{d}_i^M|$. $\qquad \square$

Lemma 8 shows that with linear precoding at the source nodes and RLNC at the intermediate nodes, the capacity region of the multi-source single-sink erasure networks as shown in Figure 3.6 is asymptotically achievable with a sufficiently large number of time extensions. Furthermore, it can be easily observed that the destination can decode all symbols by solving linear equations given in Equation (3.75).

3.4.3 Multi-source Erasure Network with Broadcast Channels

The network model given in the preceding subsections assumes that orthogonal channels are allocated for different edges, which is suitable for wireline networks with all channels physically separated. However, in wireless networks, due to the broadcast nature of wireless signals, the transmissions between different nodes are coupled by inter-link interference. One possible way to eliminate interference is to allocate orthogonal channels to different links, in which case the network can be essentially modeled by the graph shown in Figure 3.6. However, since a large number of orthogonal channels are generally required, such a method is spectrum inefficient.

The network model incorporating broadcast nature of wireless communication has been studied in [7], where each node needs to send the *same* information on all the outgoing edges. However, the messages entering the same node are still assumed to be conveyed via orthogonal channels and hence do not interfere with each other.

Theorem 2. *[7] The capacity of the multi-source single-destination networks with broadcast channels is given by*

$$
\mathcal{R}_b(G) \triangleq \left\{ \begin{array}{c} (R_i, s_i \in S) \, |0 \leq \sum_{i \in I} R_i \leq \min_{V' \subseteq (V \setminus t): s_I \subseteq V'} \bar{C}(V'), \\ \forall I \subseteq \{1, ..., k\} \end{array} \right\}, \quad (3.82)
$$

where the capacity of the edge cut, $\bar{C}(V')$, is re-defined as

$$
\bar{C}(V') = \sum_{v \in V_s'} \left(1 - \prod_{e: e \in C(V'), \text{tail}(e) = v} \epsilon_e \right) \quad (3.83)
$$

Due to the broadcast constraint, the edge cut capacity $\bar{C}(V')$ in Equation (3.83) is different from that in Equation (3.70) for networks with orthogonal channels.

With RLNC, it is observed that the broadcast erasure channel, which can be modeled by a hyperedge as illustrated in Figure 3.8(a), is equivalent to a network with deterministic orthogonal channels, as shown in Figure 3.8(b). The capacities of these deterministic links are labeled in Figure 3.8(b). Effectively, the symbols received by both receiver nodes in the broadcast channel are conveyed through an intermediate node v' via a deterministic channel of capacity $(1 - \epsilon_1)(1 - \epsilon_2)$. The symbols received by u_1 (or u_2) only are conveyed via a deterministic channel of capacity $(1 - \epsilon_1)\epsilon_2$ (or $\epsilon_1(1 - \epsilon_2)$). With RLNC, only the number of independent symbols matters for decoding. Hence, it can be verified that the deterministic network shown in Figure 3.8(b) is equivalent to the broadcast erasure channel shown in Figure 3.8(a).

The equivalence demonstrated in Figure 3.8 can be extended to general broadcast erasure channels with p receiving nodes. The corresponding deterministic network with orthogonal channels will have p links connecting to p receiving nodes and $2^p - p - 1$ intermediate nodes. Each intermediate node conveys the symbols for a unique subset with two or more receiving nodes.

Example 4. *Consider the unicast erasure network shown in Figure 3.9(a), which was constructed in [7]. The equivalent deterministic network with*

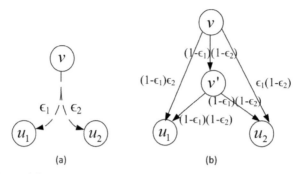

Figure 3.8   (a) The broadcast channel. (b) The equivalent network.

RLNC is shown in Figure 3.9(b). The capacity of the edge cut in Figure 3.9(a) can be evaluated according to Equation (3.83) as

$$\bar{C}(\{s, v\}) = 2 - \epsilon_1 - \epsilon_3 \epsilon_4. \tag{3.84}$$

The capacity of the corresponding cut in the equivalent network shown in Figure 3.9(b) is evaluated from Equation (3.70) by simply summing up the edge capacity as

$$
\begin{aligned}
C(\{s, v\}) &= (1 - \epsilon_1)\epsilon_2 + (1 - \epsilon_1)(1 - \epsilon_2) \\
&\quad + \epsilon_3(1 - \epsilon_4) + (1 - \epsilon_3)(1 - \epsilon_4) + (1 - \epsilon_3)\epsilon_4 \\
&= 2 - \epsilon_1 - \epsilon_3 \epsilon_4. \tag{3.85}
\end{aligned}
$$

Following from the results presented in the preceding subsections for networks with orthogonal channels, as well as the equivalence between the

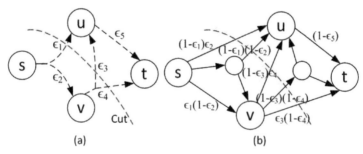

Figure 3.9   (a) An erasure network with broadcast channel. (b) The equivalent deterministic network with orthogonal channels.

broadcast erasure networks and the networks with orthogonal channels, we can conclude that the capacity for the multi-source single-sink erasure network with broadcast constraint can also be achieved by linear precoding at the source nodes together with RLNC at the intermediate nodes.

3.4.4 General Model for Wireless Erasure Networks

The general model for wireless erasure networks has been introduced in [36], which incorporates both *broadcast* at the transmitters and *interference* at the receivers. Based on this model, the total bandwidth is divided into a few orthogonal channels using multiple access methods, e.g., FDMA, TDMA or CDMA. Each node is allocated with some orthogonal channels for transmitting and receiving messages. Different messages can be broadcasted to the downstream nodes via orthogonal channels and the messages received at the common input of a node interfere with each other. The interference is modeled by finite-field sum in [36] and it is extended to general finite-field linear functions in [35].

For general model of wireless erasure networks, the capacity of a cut is given by the rank of the associated transfer matrix defined in Chapter 2 of [36]. Let $\mathbf{M}(V')$ denote the transfer matrix for the cut $C(V')$, the capacity of a *unicast* wireless erasure network is shown to be [36]:

$$R \leq \min_{V' \subseteq (V \setminus t) : s \in V'} \mathbb{E}[\text{rank}(\mathbf{M}(V'))], \tag{3.86}$$

where $\mathbb{E}[\cdot]$ is the expectation over all possible erasure events on the edge cut $C(V')$.

By extending the results to multi-source erasure networks, we can obtain the corresponding capacity region given by

$$\mathcal{R}_g(G) \triangleq \left\{ \begin{array}{c} (R_i, s_i \in S) \, | \, 0 \leq \sum_{i \in I} R_i \leq \min_{V' \subseteq (V \setminus t) : s_I \subseteq V'} \tilde{C}(V'), \\ \forall I \subseteq \{1, 2, ..., k\} \end{array} \right\}, \tag{3.87}$$

where $\tilde{C}(V') \triangleq \mathbb{E}[\mathbf{M}(V')]$.

For this general model of wireless erasure networks where the links entering the same node interfere each other, it is not yet known whether RLNC is still optimal.

3.5 Notes and Further Reading

In this chapter, we have presented a low-complexity approach to construct network codes for some classes of multi-session networks. With the proposed approach, RLNC is applied at the intermediate nodes and proper precoding is designed for the source nodes to minimize the inter-session interference and maximize the transmission rate of desired information. With RLNC at the intermediate nodes, the network is modeled by deterministic communication channels with specific decoding requirements. Therefore, the resulting problem is similar to the degree of freedom (DoF) characterization for MIMO communication systems [37, 40]. However, the existing results for MIMO systems cannot be directly used since the two systems also differ significantly. Specifically, the resulting channel matrices with RLNC are defined in a finite field, and they may be *correlated* and/or *rank-deficient*. We have utilized the similar ideas for MIMO communications to design the proper precoders in networks, e.g., rate-splitting is adopted for designing the optimal precoder of double-unicast networks in this chapter, and interference alignment is applied to design the precoder of 3-unicast networks in [15].

As the number of communication sessions in the network increases, the resulting inter-session interference with RLNC becomes more challenging to deal with. Therefore, extending the results in this chapter to large multi-session networks is highly non-trivial. One possible approach is to divide the network into a few subnetworks each containing one or two communication sessions, where intra-session network coding [41] or the proposed RLNC with source precoding [16] can be applied.

Appendix

Fact 1. *Let* $\mathbf{G} \in \mathbb{F}_q^{l \times u}$ *be a given matrix with* $\mathrm{rank}(\mathbf{G}) = r$, *and* $\mathbf{B} \in \mathbb{F}_q^{k \times l}$ *be a random matrix whose entries are uniformly generated from* $\mathbb{F}_q$. *Then, we have* $\mathrm{rank}(\mathbf{BG}) = \min(k, r)$ *with high probability as* $q \to \infty$.

Proof. Fact 1 can be shown by separately considering the two cases with $k \le r$ and $k > r$.

Case 1: $k \le r$. In this case, we can construct a submatrix $\mathbf{G}^s \in \mathbb{F}_q^{l \times k}$ from $\mathbf{G}$ by selecting its k independent columns. Then we have

$$\mathrm{rank}(\mathbf{BG}) \ge \mathrm{rank}(\mathbf{BG}^s) = k, \quad \text{as } q \to \infty, \tag{3.88}$$

where the last equality can be shown by considering the determinant of the square matrix $\mathbf{BG}^s \in \mathbb{F}_q^{k \times k}$. Specifically, as $\mathbf{B}$ is randomly generated, the determinant of $\mathbf{B}\mathbf{G}^s$ is a polynomial on the entries of $\mathbf{B}$. If this polynomial is not identically zero, then with high probability as $q \to \infty$, the determinant of $\mathbf{B}\mathbf{G}^s$ is nonzero, and hence $\mathrm{rank}(\mathbf{B}\mathbf{G}^s) = k$. To show that the polynomial is not identically zero, it is sufficient to choose one particular matrix for $\mathbf{B}$, e.g., $\mathbf{B} = (\mathbf{G}^s)^T$, so that $\mathrm{rank}(\mathbf{B}\mathbf{G}^s) = \mathrm{rank}((\mathbf{G}^s)^T \mathbf{G}^s) = k$. Therefore, the determinant of $\mathbf{B}\mathbf{G}^s$ is not identically zero and we have $\mathrm{rank}(\mathbf{B}\mathbf{G}^s) = k$ with high probability as $q \to \infty$.

Besides, by considering the matrix dimensions of $\mathbf{B}\mathbf{G}$, we obviously have $\mathrm{rank}(\mathbf{B}\mathbf{G}) \le k$. Together with Equation (3.88), we have the desired result: $\mathrm{rank}(\mathbf{B}\mathbf{G}) = k = \min\{k, r\}$ for $k \le r$.

Case 2: $k > r$. In this case, $\mathbf{B}$ can be decomposed as $\mathbf{B} = \begin{bmatrix} \mathbf{B}_1 \\ \mathbf{B}_2 \end{bmatrix}$, where $\mathbf{B}_1 \in \mathbb{F}_q^{r \times l}$ and $\mathbf{B}_2 \in \mathbb{F}_q^{(k-r) \times l}$. Following similar arguments to those in Case 1, we can show that $\mathrm{rank}(\mathbf{B}\mathbf{G}) \ge \mathrm{rank}(\mathbf{B}_1\mathbf{G}) = r$. Together with the fact that $\mathrm{rank}(\mathbf{B}\mathbf{G}) \le \mathrm{rank}(\mathbf{G}) = r$, we have $\mathrm{rank}(\mathbf{B}\mathbf{G}) = \min\{k, r\}$ for $k > r$ as well.

This completes the proof of Fact 1. $\qquad\qquad\qquad\qquad\qquad\square$

Fact 2. *Let* $\mathbf{G}_1 \in \mathbb{F}_q^{l \times u_1}$ *and* $\mathbf{G}_2 \in \mathbb{F}_q^{l \times u_2}$ *be two given matrices with* $\mathrm{rank}(\mathbf{G}_1) = r_1$, $\mathrm{rank}(\mathbf{G}_2) = r_2$ *and* $\mathrm{rank}\left(\begin{bmatrix} \mathbf{G}_1 & \mathbf{G}_2 \end{bmatrix}\right) = r_{12}$. *Let* $\mathbf{B} \in \mathbb{F}_q^{u_1 \times k}, k \le r_1$, *be a random matrix whose entries are uniformly generated from* $\mathbb{F}_q$. *Then we have* $\mathrm{rank}\left(\begin{bmatrix} \mathbf{G}_1\mathbf{B} & \mathbf{G}_2 \end{bmatrix}\right) = \min\{r_{12}, k + r_2\}$ *with high probability as* $q \to \infty$.

Proof. Fact 2 can be shown by separately considering the two cases with $k \le r_{12} - r_2$ and $k > r_{12} - r_2$.

Case 1: $k \le r_{12} - r_2$. Since $\mathrm{rank}\left(\begin{bmatrix} \mathbf{G}_1\mathbf{B} & \mathbf{G}_2 \end{bmatrix}\right) \le \mathrm{rank}(\mathbf{G}_1\mathbf{B}) + \mathrm{rank}(\mathbf{G}_2) = k + r_2$, we only need to show that $\mathrm{rank}\left(\begin{bmatrix} \mathbf{G}_1\mathbf{B} & \mathbf{G}_2 \end{bmatrix}\right) \ge k + r_2$. To this end, let $\mathbf{G}_2^s$ be a submatrix of $\mathbf{G}_2$ consisting of its r_2 independent columns. Since $\mathrm{rank}(\mathbf{G}_2) = r_2$, the columns in $\mathbf{G}_2^s$ span the same subspace as those in $\mathbf{G}_2$. Thus, we have $\mathrm{rank}\left(\begin{bmatrix} \mathbf{G}_1 & \mathbf{G}_2^s \end{bmatrix}\right) = \mathrm{rank}\left(\begin{bmatrix} \mathbf{G}_1 & \mathbf{G}_2 \end{bmatrix}\right) = r_{12}$. Let $\mathbf{U}$ be a random matrix uniformly generated from $\mathbb{F}_q^{(k+r_2) \times l}$. According to Fact 1, we have

$$\mathrm{rank}\left(\mathbf{U}\begin{bmatrix} \mathbf{G}_1 & \mathbf{G}_2^s \end{bmatrix}\right) = \min\{k + r_2, r_{12}\} = k + r_2. \qquad (3.89)$$

Let $\tilde{\mathbf{G}}_1 = \mathbf{U}\mathbf{G}_1$ and $\tilde{\mathbf{G}}_2^s = \mathbf{U}\mathbf{G}_2^s$. Then we have

$$\mathrm{rank}\left(\begin{bmatrix} \mathbf{G}_1\mathbf{B} & \mathbf{G}_2 \end{bmatrix}\right) \ge \mathrm{rank}\left(\begin{bmatrix} \mathbf{G}_1\mathbf{B} & \mathbf{G}_2^s \end{bmatrix}\right) \ge \mathrm{rank}\left(\mathbf{U}\begin{bmatrix} \mathbf{G}_1\mathbf{B} & \mathbf{G}_2^s \end{bmatrix}\right)$$
$$= \mathrm{rank}\left(\begin{bmatrix} \tilde{\mathbf{G}}_1\mathbf{B} & \tilde{\mathbf{G}}_2^s \end{bmatrix}\right) = k + r_2, \qquad (3.90)$$

where the last equality can be shown by verifying that the determinant of the square matrix $\begin{bmatrix} \tilde{\mathbf{G}}_1 \mathbf{B} & \tilde{\mathbf{G}}_2^s \end{bmatrix}$ as a polynomial of entries in $\mathbf{B}$ is not identically zero. To this end, we can choose one particular $\mathbf{B}$ such that $\text{rank}\left(\begin{bmatrix} \tilde{\mathbf{G}}_1 \mathbf{B} & \tilde{\mathbf{G}}_2^s \end{bmatrix}\right) = k + r_2$. According to Equation (3.89), there are at least k independent columns in $\tilde{\mathbf{G}}_1$ that are not in the subspace spanned by $\tilde{\mathbf{G}}_2^s$. Denote the indices of these columns by $i_1, i_2, ..., i_k$. If $\mathbf{B}$ is chosen such that the jth column is a unit vector with non-zero value only at the i_jth entry, $j = 1, ..., k$, then $\tilde{\mathbf{G}}_1 \mathbf{B}$ is a submatrix of $\tilde{\mathbf{G}}_1$ with columns not in the subspace spanned by $\tilde{\mathbf{G}}_2^s$; hence $\begin{bmatrix} \tilde{\mathbf{G}}_1 \mathbf{B} & \tilde{\mathbf{G}}_2^s \end{bmatrix}$ is a full-rank matrix. Therefore, we can conclude that the determinant of $\begin{bmatrix} \tilde{\mathbf{G}}_1 \mathbf{B} & \tilde{\mathbf{G}}_2^s \end{bmatrix}$ is not an identically zero polynomial and $\text{rank}\left(\begin{bmatrix} \tilde{\mathbf{G}}_1 \mathbf{B} & \tilde{\mathbf{G}}_2^s \end{bmatrix}\right) = k + r_2$ with high probability as $q \to \infty$.

Case 2: $k > r_{12} - r_2$. Since $\text{rank}\left(\begin{bmatrix} \mathbf{G}_1 \mathbf{B} & \mathbf{G}_2 \end{bmatrix}\right) \leq \text{rank}\left(\begin{bmatrix} \mathbf{G}_1 & \mathbf{G}_2 \end{bmatrix}\right) = r_{12}$, we only need to show that $\text{rank}\left(\begin{bmatrix} \mathbf{G}_1 \mathbf{B} & \mathbf{G}_2 \end{bmatrix}\right) \geq r_{12}$. Let $\mathbf{V} \in \mathbb{F}_q^{(r_{12} \times l)}$ be a random matrix with entries uniformly generated from $\mathbb{F}_q$, and let $\bar{\mathbf{G}}_1 = \mathbf{V}\mathbf{G}_1$ and $\bar{\mathbf{G}}_2 = \mathbf{V}\mathbf{G}_2$. According to Fact 1, we have $\text{rank}(\bar{\mathbf{G}}_1) = r_1$, $\text{rank}(\bar{\mathbf{G}}_2) = r_2$, and $\text{rank}\left(\begin{bmatrix} \bar{\mathbf{G}}_1 & \bar{\mathbf{G}}_2 \end{bmatrix}\right) = r_{12}$. Therefore, we can construct a matrix $\bar{\mathbf{G}}_2^{s_1}$ by choosing $(r_{12} - r_1)$ independent columns from $\bar{\mathbf{G}}_2$ that are not in the subspace spanned by $\bar{\mathbf{G}}_1$. Furthermore, let $\bar{\mathbf{G}}_2^{s_2}$ be a submatrix of $\bar{\mathbf{G}}_2$ consisting of $(r_1 - k)$ columns that are not in the subspace spanned by $\bar{\mathbf{G}}_2^{s_1}$. Such a choice is feasible since $(r_{12} - r_1) + (r_1 - k) = r_{12} - k \leq r_2$. Then, it follows that $\text{rank}\left(\begin{bmatrix} \bar{\mathbf{G}}_1 & \bar{\mathbf{G}}_2^{s_1} & \bar{\mathbf{G}}_2^{s_2} \end{bmatrix}\right) = r_{12}$. Following similar arguments to those for case 1, we can show that $\begin{bmatrix} \bar{\mathbf{G}}_1 \mathbf{B} & \bar{\mathbf{G}}_2^{s_1} & \bar{\mathbf{G}}_2^{s_2} \end{bmatrix}$ is a full-rank square matrix with high probability as $q \to \infty$. Hence, we have the desired result:

$$\text{rank}\left(\begin{bmatrix} \mathbf{G}_1 \mathbf{B} & \mathbf{G}_2 \end{bmatrix}\right) \geq \text{rank}\left(\begin{bmatrix} \bar{\mathbf{G}}_1 \mathbf{B} & \bar{\mathbf{G}}_2 \end{bmatrix}\right) \geq \text{rank}\left(\begin{bmatrix} \bar{\mathbf{G}}_1 \mathbf{B} & \bar{\mathbf{G}}_2^{s_1} & \bar{\mathbf{G}}_2^{s_2} \end{bmatrix}\right) = r_{12}.$$

$\square$

References

[1] Yeung, R. W., and Zhang, Z. (1999). Distributed source coding for satellite communications. *IEEE Trans. Inform. Theory* 45, 1111–1120.

[2] Wu, Y., Chou, P. A., and Kung, S. Y. (2005). Minimum-energy multicast in mobile ad hoc networks using network coding. *IEEE Trans. Commun.* 53, 1906–1918.

[3] Katti, S., Rahul, H., Hu, W., Katabi, D., Medard, M., and Crowcroft, J. (2008). XORs in the air: practical wireless network coding. *IEEE/ACM Trans. On Netw.* 16, 497–510.

[4] Gkantsidis, C., and Rodriguez, P. (2005). "Network coding for large scale content distribution," in *Proceedings of the IEEE 24th Annual Joint Conference of the IEEE Computer and Communications Societies (INFOCOM)*, Miami, FL, 2235–2245.

[5] Dimakis, A. G., Godfrey, P. B., Wu, Y., Wainwright, M. J., and Ramchandran, K. (2010). Network coding for distributed storage systems. *IEEE Trans. Inform. Theory* 56, 4539–4551.

[6] Li, S.-Y. R., Yeung, W., and Cai, N. (2003). Linear network coding. *IEEE Trans. Inform. Theory* 49, 371–381.

[7] Dana, A. F., Gowaikar, R., Palanki, R., Hassibi, B., and Effros, M. (2006). Capacity of wireless erasure networks. *IEEE Trans. Inform. Theory* 52, 789–804.

[8] Ho, T., Koetter, R., Medard, M., Effros, M., Shi, J., and Karger, D. (2006). A random linear network coding approach to multicast. *IEEE Trans. Inform. Theory* 52, 4413–4430.

[9] Dougherty, R., Freiling, C., and Zeger, K. (2005). Insufficiency of linear coding in network information flow. *IEEE Trans. Inform. Theory* 51, 2745–2759.

[10] Traskov, D., Ratnakar, N., Lun, D., Koetter, R., and Medard, M. (2006). "Newtork coding for multiple unicasts: An approach based on linear optimization," in *Proceedings of the IEEE International Symposium on Information Theory,* Seattle, WA, 1758–1762.

[11] Wang, C. C., and Shroff, N. B. (2010). Pairwise intersession network coding on directed networks. *IEEE Trans. Inform. Theory* 56, 3879–3900.

[12] Heindlmaier, M., Lun, D. S., Traskov, D., and Medard, M. (2011). "Wireless inter-session network coding – an approach using virtual multicasts," in *Proceedings of the IEEE International Conference on Communications (ICC)*, Kyoto, 1–5.

[13] Erez, E., and Feder, M. (2003). "Capacity region and network codes for two receivers multicast with private and common data," in *Proceedings of the Workshop on Coding, Cryptography and Combinatorics*, Huangshan.

[14] Erez, E., and Feder, M. (2009). Improving the multicommodity flow rate with network codes for two sources. *IEEE J. Select. Areas Comm.* 27, 814–824.

[15] Das, A., Vishwanath, S., Jafar, S., and Markopoulou, A. (2010). "Network coding for multiple unicasts: An interference alignment approach,"

in *IEEE International Symposium on Information Theory*, Austin, TX, 1878–1882.

[16] Xu, X., Zeng, Y., Guan, Y. L., and Ho, T. (2014). An achievable region for double-unicast networks with linear network coding. *IEEE Trans. Commun.* 62, 3621–3630.

[17] Xu, X., Zeng, Y., and Guan, Y. L. (2015). "Multi-source erasure networks with source precoding and random linear network coding," in *IEEE International Conference on Information, Communications and Signal Processing*, Singapore, December 2015.

[18] Huang, S., and Ramamoorthy, A. (2013). An achievable region for the double unicast problem based on a minimum cut analysis. *IEEE Trans. Commun.* 61, 2890–2899.

[19] Li, Z., and Li, B. (2004). "Network coding: The case of multiple unicast sessions," in *Proceedings of the 42nd Annual Allerton Conference on Communication, Control, and Computing*, October 2004.

[20] Dougherty, R., and Zeger, K. (2006). Nonreversibility and equivalent constructions of multiple-unicast networks. *IEEE Trans. Inform. Theory* 52, 5067–5077.

[21] Shenvi, S., and Dey, B. K. (2010). "A simple necessary and sufficient condition for the double unicast problem," in *Proceedings of theIEEE International Conference on Communications (ICC)*, Cape Town, 1–5.

[22] Wang, C. C., and Shroff, N. B. (2007). "Beyond the butterfly a graph-theoretic characterization of the feasibility of network coding with two simple unicast sessions," in *Proceedings of the IEEE International Symposium on Information Theory*, Nice, 121–125.

[23] Kamath, S., Tse, D., and Wang, C. C. (2014). "Two-unicast is hard," in *Proceedinhs of the IEEE International Symposium on Information Theory*, Honolulu, Hawaii, Jun. 2014.

[24] Gamal, A. E., and Costa, M. H. M. (1982). The capacity region of a class of deterministic interference channels. *IEEE Trans. Inform. Theory* 28, 343–346.

[25] Han, T., and Kobayashi, K. (1981). A new achievable region for the interference channel. *IEEE Trans. Inform. Theory* 27, 49–60.

[26] Bernstein, D. S. (2009). *Matrix Mathematics: Theory, Facts and Formulas*. Princeton, NJ: Princeton University Press.

[27] Zeng, W., Cadambe, V., and Medard, M. (2012). "An edge reduction lemma for linear network coding and an application to two-unicast networks," in *Proceedings of the 50th Annual Allerton Conference on Communication, Control, and Computing*, October 2012.

[28] Xu, X., and Guan, Y. L. (2013). "Joint routing and random network coding for multi-session networks," in *Proceedings of the 19th IEEE International Conference on Networks*, Singapore, October 2013.

[29] Appuswamy, R., Franceschetti, M., Karamchandani, N., and Zeger, K. (2011). Network coding for computing: Cut-set bounds. *IEEE Trans. Inform. Theory* 57, 1015–1030.

[30] Lehman, A. R., and Lehman, E. (2004). "Complexity classification of network information flow problems," in *Proceedings of the Symposium on Discrete Algorithms*, New Orleans, LA, 142–150.

[31] Dikaliotis, T. K., Ho, T., Jaggi, S., Vyetrenko, S., Yao, H., Effros, M., et al. (2011). Mutliple-access network information-flow and correction codes. *IEEE Trans. Inform. Theory* 57, 1067–1079.

[32] Julian, D. (2002). "Erasure networks," in *Proceedings of the IEEE International Symposium on Information Theory*, Lausanne.

[33] Smith, B., and Vishwanath, S. (2007). "Unicast transmission over multiple access erasure networks: capacity and duality," in *Proceedings of the IEEE Information Theory Workshop*, Tahoe City, CA, 331–336.

[34] Gowda, K. T., and Sumei, S. (2008). "Multicast capacity of multiple access erasure networks," in *Proceedings of the International Symposium on Information Theory and Its Applications*, Auckland.

[35] Avestimehr, S., Diggavi, S. N., and Tse, D. (2007). "Wireless network information flow," in *Proceedings of the 45th Annual Allerton Conference on Communication, Control and Computing*, Monticello, IL.

[36] Smith, B. M. (2008). *Capacities of Erasure Networks*. Ph.D. Dissertation, 2008. Available at: http://utlinc.org/images/9/9a/Bms_proposal.pdf

[37] Cadambe, V. R., and Jafar, S. A. (2008). Interference alignment and degrees of freedom of the k-user interference channel. *IEEE Trans. Inform. Theory* 54, 3425–3441.

[38] Gamal, A. E., and Kim, Y. H. (2012). *Network Information Theory*. Cambridge: Cambridge University Press.

[39] Lun, D. S., Medard, M., Koetter, R., and Effros, M. (2008). On coding for reliable communication over packet networks. *Phys. Commun.* 1, 3–20.

[40] Zeng, Y., Xu, X., Guan, Y. L., Gunawan, E. and Wang, C. (2014). Degrees of freedom of the three-user rank-deficient MIMO interference channel. *IEEE Trans. Wirel. Commun.* 13, 4179–4192.

[41] Ho, T., and Viswanathan, H. (2009). Dynamic algorithms for multicast with intra-session network coding. *IEEE Trans. Inform. Theory* 55, 797–815.

4

Decoding Scheduling for Low-Density Parity-Check Codes

Huang-Chang Lee[1], Yen-Ming Chen[2] and Yeong-Luh Ueng[3]

[1]Department of Electrical Engineering, Chang Gung University,
Taoyuan City, Taiwan
[2]Institute of Communications Engineering, National Sun Yat-sen University,
Kaohsiung, Taiwan
[3]Department of Electrical Engineering and the Institute of Communications
Engineering, National Tsing Hua University, Hsinchu, Taiwan

4.1 Introduction

Low-density parity-check (LDPC) [1] codes are capable of proving near-capacity performances, and have been adopted in the standard of the fifth-generation mobile communications (5G) [2]. When the iterative belief propagation (BP) algorithm is applied for LDPC decoding, the convergence speed of the error rate in the waterfall region is affected by decoding schedules.

The most straightforward decoding schedule is the flooding schedule, where all decoding messages are propagated simultaneously. In order to accelerate the convergence speed, serial schedules can be used, and the decoding messages can be generated according to the information produced during the same iteration. The schedules presented in [3–8] achieve similar performances, and are called standard sequential scheduling (SSS) in [9]. Layered BP (LBP) [4, 5] and shuffled decoding [3] are commonly used serial schedules which, respectively, arrange the decoding process in the order of rows or columns of the parity-check matrix. These hardware-friendly fixed schedules, many sophisticated decoding circuits have been implemented [10, 11].

Based on the concept of providing the maximum increase in the mutual information for the messages exchanged in the Tanner graph, the maximum mutual information increase (M^2I^2)-based algorithm was proposed in [12]. The M^2I^2-based algorithm first measures the increase in the mutual information that can be provided after each update to the messages. Accordingly, the decoding schedule is then arranged by sequentially predicting the updates that provide the maximum increase in mutual information. Not only can the prediction be performed by using the mutual information for the messages to be updated in the next decoding stage, but it can also be further refined using the mutual information for the messages to be updated in the subsequent decoding stages. The schedule will then be determined so as to supply long-term benefits and provide a faster convergence speed. Since the decoding sequence arranged using the M^2I^2-based algorithm can be pre-determined and fixed, the only increase in hardware implementation complexity will be the storage of the update sequence. The simulation results show that, following the schedule arranged using the M^2I^2-based algorithm, the number of iterations can be reduced by over 60% compared to that of the conventional flooding schedule.

In addition to the convergence speed, the error-floor performance of LDPC codes is also critical to applications such as storage systems and optical communications, where extremely low error-floor values are demanded. Due to the use of sub-optimum iterative BP algorithm, trapping sets [13] can easily be triggered, which are regarded as the major cause of the error floor. Many studies have also shown that once trapping sets are triggered, their effects can be mitigated by using modified decoding strategies [14].

In addition, the concept of *schedule diversity* is introduced in [12]. It is demonstrated that the effect of trapping sets can be mitigated by using scheduling techniques, where a lower error floor can be achieved without requiring any knowledge of the trapping sets. For a single received frame, the values of the decoding messages exchanged using different schedules will not be exactly equal, and diverse decoding results can be obtained. Conventionally, only a single decoding schedule is adopted, and a relatively low error rate can be achieved in the high SNR (signal-to-noise ratio) region for a well-constructed code, which implies that the probability of falling into trapping sets is low. In contrast, if multiple decoding attempts based on multiple distinct decoding schedules can be applied to a single received frame, schedule diversity can be achieved, and the probability of falling into trapping sets would be even lower. Simulation results show that when the proposed schedule diversity is implemented, a more significant improvement in error-floor performance can be achieved, when compared to that of the backtracking algorithm reported in [15].

4.2 Belief Propagation Decoding for LDPC Codes

For a binary LDPC code described using a bipartite graph consisting of N variable nodes and M check nodes, the j-th variable node can be denoted as v_j, and the i-th check node can be denoted as c_i. When the BP algorithm is applied to LDPC codes, variable-to-check (V2C) and check-to-variable (C2V) messages are iteratively updated during the decoding process. The V2C message propagated from variable node v_j to check node c_i is denoted as $L_{v_j \rightarrow c_i}$. Similarly, the C2V message propagated from check node c_i to variable node v_j is denoted as $m_{c_i \rightarrow v_j}$. As the decoding scheme progresses, the *a posteriori* log-likelihood ratio (LLR) of variable node v_j, denoted as L_{v_j}, will be gradually refined by the C2V messages. The values of $L_{v_j \rightarrow c_i}$, $m_{c_i \rightarrow v_j}$, and L_{v_j} can be respectively generated using Equations (4.1–4.3), as follows:

$$L_{v_j \rightarrow c_i} = \sum_{c_a \in \mathcal{N}(v_j) \backslash c_i} m_{c_a \rightarrow v_j} + \Lambda_{v_j}, \tag{4.1}$$

$$m_{c_i \rightarrow v_j} = 2 \times \tanh^{-1} \left(\prod_{v_b \in \mathcal{N}(c_i) \backslash v_j} \tanh \left(\frac{L_{v_b \rightarrow c_i}}{2} \right) \right), \tag{4.2}$$

$$L_{v_j} = \sum_{c_a \in \mathcal{N}(v_j)} L_{c_a \rightarrow v_j} + \Lambda_{v_j}, \tag{4.3}$$

where Λ_{v_j} represents the intrinsic LLR of v_j, which is also known as the channel value, and $\mathcal{N}(n_1) \backslash n_2$ denotes the set which consists of the neighbors of node n_1, excluding node n_2, where nodes n_1 and n_2 can be either a variable node or a check node.

4.3 Fixed Schedules

Equations (4.1) and (4.2) describe the rules for generating the messages that will be exchanged between neighboring nodes. For a reliable hard-decision decoding result, several iterations are required to obtain converged LLR values L_{v_j} using Equation (4.3). It can be observed that the operations for generating C2V and V2C messages can be ordered in different schedules.

In the following, the widely used conventional parallel and serial fixed schedules will be introduced.

4.3.1 The Flooding Schedule

Flooding (FL) is the most straightforward scheduling strategy. It is also referred to as the two-phase message passing (TPMP), where all V2C messages are updated in the first phase using Equation (4.1), and then all C2V messages are updated in the second phase using Equation (4.2). Following the FL schedule, a single node can only generate new messages after it has been updated by all its neighboring nodes.

4.3.2 Standard Sequential Schedules

If the decoding messages can be sequentially updated, the previously updated information can be used by the following updates, and more up-to-date decoding messages can be generated within the same iteration. It can be expected that using these more up-to-date messages, the required number of iterations for the same decoding performance can be reduced, and therefore a faster convergence speed in error performance can be achieved.

Layered BP (LBP) [4, 5] and shuffled decoding [3] are two commonly used serial schedules. The decoding messages are updated in the order of rows and columns of the parity-check matrices, respectively, for the LBP and shuffled schedules. The demand for reducing the number of decoding iterations continuously drives the design of decoding schedules. Code structures were also considered in the design of fixed decoding schedules as reported in [16] and [17], where it was suggested that the check nodes with higher degrees could be updated more frequently [16], or that they could be updated first in each of the iterations [17]. It is shown that these fixed sequential schedules improves the convergence speed by about twice that of the flooding schedule [18, 19].

4.3.3 Decoding Schedules for Faster Convergence

For iterative BP algorithms, the variable-node operation in Equation (4.1) and the check-node operation in Equation (4.2) are edge-wise operations, and can be executed by following a variety of schedules. In this section, an M^2I^2-based algorithm that arranges the edge-wise operations for fast convergence is proposed. Considering hardware-friendly implementations, single-edge protograph-based LDPC codes, which have been adopted in many standards

[20–22], are adopted. As a result, for the design of the M^2I^2-based algorithm, we focus on single-edge protograph-based LDPC codes.

4.3.4 Protograph-based LDPC Codes

A binary LDPC code can be described using a Tanner graph consisting of N variable nodes and M check nodes, which can be obtained form the structure of the M-by-N parity-check matrix. For the code construction, structured LDPC codes, such as protograph-based LDPC codes [23], are preferred for practical implementations. The protograph-based LDPC codes can be constructed based on a blueprint, or a protograph, which can be represented by using an M_p-by-N_p base matrix **B**. The base matrix contains the information for the "type" of the variable nodes, the check nodes, and the connective edges. The entry at the i-th row and j-th column, i.e., $b_{i,j}$, denotes the number of type-(i, j) edges between the type-i check nodes and the type-j variable nodes in the protograph. In the beginning, the codes are constructed by "lifting" the protograph z times, where z denotes the lifting factor. The type-(i, j) edge in the protograph with a non-zero $b_{i,j}$ value expands to a z-by-z sub-matrix $\mathbf{h}_{i,j}$ whose row weight and column weight are both equal to $b_{i,j}$, and each of the zero elements in the base matrix expands to a z-by-z all-zero matrix. Accordingly, the connections for the same type of nodes in the different copies of the protograph are permutated. After the "copy-and-permutate" operations, an M-by-N parity-check matrix **H**, where $N = N_p \cdot z$, and $M = M_p \cdot z$, can be constructed. An example of the base matrix **B** considering $M_p = 3$ and $N_p = 4$ is given by

$$\mathbf{B} = \begin{bmatrix} 1 & 1 & 1 & 0 \\ 1 & 1 & 1 & 1 \\ 0 & 1 & 0 & 1 \end{bmatrix}, \mathbf{H} = \begin{bmatrix} \mathbf{h}_{1,1} & \mathbf{h}_{1,2} & \mathbf{h}_{1,3} & \mathbf{0}_{z \times z} \\ \mathbf{h}_{2,1} & \mathbf{h}_{2,2} & \mathbf{h}_{2,3} & \mathbf{h}_{2,4} \\ \mathbf{0}_{z \times z} & \mathbf{h}_{3,2} & \mathbf{0}_{z \times z} & \mathbf{h}_{3,4} \end{bmatrix}, \quad (4.4)$$

where $b_{i,j} \in \{0, 1\}$ for $i = 1, \cdots, M_p$ and $j = 1, \cdots, N_p$. Accordingly, all corresponding edges in the protograph are single connections. With a lifting factor z of 3, a complete Tanner graph which consists of $M = 9$ check nodes and $N = 12$ variable nodes, is illustrated in Figure 4.1.

4.3.5 Protograph-based Edge-wise Schedule

The overall sequential decoding process can be divided into several stages, and only part of the decoding messages would be updated in each stage. For

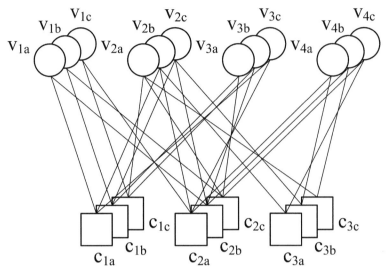

Figure 4.1   Tanner graph for the constructed low-density parity-check (LDPC) code considering the base matrix **B** given in Equation (4.4).

a single-edge protograph-based LDPC code, messages to be updated in each stage can be indicated by the index of the corresponding protograph edge. Consequently, the decoding process can be described by using a sequence **S**, whose elements represent the indices of the protograph edges. As the decoding process progresses, the decoding messages are updated following the order of the sequence **S**. When the LBP schedule is adopted, the decoding process is arranged based on the order of the check nodes. For the LDPC codes constructed based on the protograph given in Figure 4.1, following the LBP schedule, the C2V messages which correspond to the edges connected to the type-1 check nodes will be updated first. Since the type-1 check nodes are connected to the type-1, the type-2, and the type-3 variable nodes, the first three elements of the sequence **S** are $S(1) = (1,1)$, $S(2) = (1,2)$, and $S(3) = (1,3)$. In the first stage of the decoding process, using Equation (4.2), the C2V messages $R_{1a \to 1a}$, $R_{1b \to 1b}$, and $R_{1c \to 1c}$ are updated following the order of sequence **S**. The updates to the messages generated from the type-1 check nodes are completed in the third stage. The updates to the messages generated from the type-2 check nodes begin in the fourth decoding stage, where $S(4) = (2,1)$. Similarly, the updates to the messages for the type-2 check nodes are completed in the 7-th stage, which is indicated by $S(7) = (2,4)$.

For the protograph-based edge-wise schedule, the decoding sequence **S** does not necessarily follow the order of the edge index. Consequently, it is not guaranteed that all C2V messages have the same number of updates. A counter $u_{i,j}$ is then introduced to denote the number of times that the C2V messages which correspond to the permutation sub-matrix $\mathbf{h}_{i,j}$ have been updated when $b_{i,j} \neq 0$. If $b_{i,j} = 0$, $u_{i,j}$ is always zero. As a result, for a protograph with a total number of edges E_B, the number of iterations ℓ is calculated as $\ell = \left(\sum_{i=1}^{M_p} \sum_{j=1}^{N_p} u_{i,j} \right) / E_B$. Based on the definition of ℓ, the performance and complexity of the protograph-based edge-wise schedule can be compared with those of conventional decoding schedules.

4.3.6 The M^2I^2-based Algorithm

The M^2I^2-based algorithm is aimed at designing a protograph-based edge-wise schedule from the perspective of mutual information to accelerate the convergence of the decoding process, and hence, reducing the average number of iterations required to achieve a given error-rate performance. The design of the schedule is arranged based on the edge in the protograph.

Assume that the first s elements in sequence **S** have been determined, and the M^2I^2-based algorithm is preparing to determine the $(s + 1)$-th element, i.e., $S(s + 1)$. In addition, we also assume that the C2V messages associated with type-(i, j) edges have been updated $u_{i,j}$ times during the first s decoding stages, and the corresponding C2V mutual information, i.e., the mutual information for the C2V messages, is denoted as $I_{E,C}^{u_{i,j}}(i, j)$. Since the updated messages in the $(s + 1)$-th decoding stage have not yet been determined, the predictions of the mutual information for the $(s + 1)$-th stage are calculated according to the pre-computations, rather than the actual updates. Let $I_{E,C}^{p}(i, j)$ denote the predicted C2V mutual information, which is provided by the pre-computations of the C2V messages associated with $\mathbf{h}_{i,j}$. The pre-computations are executed for all protograph edges, and the increase can be measured by

$$r(i, j) = I_{E,C}^{p}(i, j) - I_{E,C}^{u_{i,j}}(i, j). \tag{4.5}$$

Using the increase defined in Equation (4.5), the protograph edge with the maximum increase can be identified from:

$$r(i^*, j^*) \equiv \max \left\{ r(i, j) \,|\, i = 1, \cdots, M_p, j = 1, \cdots, N_p, b_{ij} \neq 0 \right\}. \tag{4.6}$$

The index values (i^*, j^*) will then be stored as the $(s + 1)$-th element of **S**, i.e., $S(s + 1) = (i^*, j^*)$. By following the order of **S**, the C2V messages

$R_{i^* \to j^*}$, which are associated with $\mathbf{h}_{i^*,j^*}$, will be updated in the $(s+1)$-th decoding stage, and hence, the predicted mutual information for the (i^*, j^*)-edge, i.e., $I_{E,C}^p(i^*, j^*)$, can be used in the $(s+1)$-th decoding stage to replace that obtained by the actual update, i.e., $I_{E,C}^{u_{i^*j^*}+1}(i^*, j^*) = I_{E,C}^p(i^*, j^*)$.

Similarly, the V2C mutual information, i.e., the mutual information for the V2C messages, can also be evaluated. If we let $I_{E,V}(i, j)$ denote the V2C mutual information provided by the updated V2C messages associated with the protograph edge (i, j), the mutual information value can be calculated using the formulas provided in [24]. Consequently, the mutual information for both the C2V and the V2C messages can be iteratively evaluated for each decoding stage. The detailed procedures of the proposed M^2I^2-based algorithm for determining $\mathbf{S}$ is described in Algorithm 1:.

In the initial phase, the value of the sequence index is set as $s = 1$, and all the update counters are set as $u_{i,j} = 0$, and the C2V mutual information associated with each edge is set as zero, i.e., $I_{E,C}^0(i, j) = 0$. The mutual information for the channel messages to type-j variable nodes is $I_{ch}(j) = J\left(\sqrt{8R(E_b/N_0)}\right)$, where

$$J(\sigma) = 1 - \int_{-\infty}^{\infty} \frac{1}{\sqrt{2\pi\sigma^2}} e^{-\frac{(y-\sigma^2/2)^2}{2\sigma^2}} \cdot \log_2(1 + e^{-y}) dy. \qquad (4.7)$$

Since $J(\cdot)$ is monotonic, the inverse function $J^{-1}(\cdot)$ exists [25]. The initial V2C mutual information $I_{E,V}(i, j)$ sent from type-j variable nodes to the neighboring type-i check nodes is equal to $I_{ch}(j)$.

To determine $S(s)$, the C2V mutual information provided by the pre-computations of the C2V messages associated with $\mathbf{h}_{i,j}$ can be evaluated as:

$$I_{E,C}^p(i, j) = 1 - J\left(\sqrt{\sum_{v_b \in \mathcal{N}(c_i)\backslash v_j} (J^{-1}(1 - I_{E,V}(i, b)))^2}\right). \qquad (4.8)$$

Accordingly, the predicted versions of the C2V mutual information and the corresponding increases for all protograph edges can be calculated. These increases are compared based on Equation (4.6), and the index (i^*, j^*) can be identified and hence stored as the s-th element of $\mathbf{S}$. As a result, the edges that will be processed in the s-th decoding stage have now been determined. As shown in lines 5 and 6 of Algorithm 1:, the C2V mutual information provided in the s-th decoding stage is equal to the prediction for edge (i^*, j^*), and the value of the update counter u_{i^*,j^*} is increased by one.

Based on the new $I_{E,C}^{u_{i^*},j^*}$ value, the V2C mutual information $I_{E,V}(i,j^*)$ value propagated from type-j^* variable nodes to the neighboring type-i check nodes can be evaluated in line 7 of Algorithm 1: as:

$$I_{E,V}(i,j^*) = J\left(\sqrt{\sum_{c_a \in \mathcal{N}(v_{j^*})\backslash c_i}(J^{-1}(I_{E,C}^{u_a,j^*}(a,j^*)))^2 + (J^{-1}(I_{ch}(j^*)))^2}\right). \quad (4.9)$$

It can be observed from Equations (4.8) and (4.9) that the mutual information for both the C2V and V2C messages is iteratively renewed by each other, and their mutual information values are gradually increased. The arrangement process progresses until the value of the cumulative mutual information (CMI) for each variable node converges to one, where the CMI for type-j variable nodes can be evaluated as:

$$I_{CMI}(j) = J\left(\sqrt{\sum_{c_a \in \mathcal{N}(v_j)}(J^{-1}(I_{E,C}(a,j)))^2 + (J^{-1}(I_{ch}(j)))^2}\right). \quad (4.10)$$

The criterion adopted in the M^2I^2-based algorithm is to maximize the increase in the C2V mutual information based on Equation (4.6). Since

---

Algorithm 1: The proposed M^2I^2-based schedule arrangement.

Initialization:

   $s = 1$, $u_{i,j} = 0$ and $I_{E,C}^0(i,j) = 0$ for all (i,j).

   $I_{ch}(j) = J\left(\sqrt{8RE_b/N_0}\right)$ for all j,

   where $E_b/N_0 > (E_b/N_0)_0$.

   $I_{E,V}(i,j) = I_{ch}(j)$ for $c_i \in \mathcal{N}(v_j)$.

1: **for**($i = 1, \cdots, M_p$, $j = 1, \cdots, N_p$, and $b_{i,j} \neq 0$)

2:   Generate $I_{E,C}^p(i,j)$ using Equation (4.8).

3:   Generate $r(i,j) = I_{E,C}^p(i,j) - I_{E,C}^{u_{ij}}(i,j)$.

4: **end for**

5: Determine (i^*,j^*) based on Equation (4.6), $S(s) = (i^*,j^*)$.

6: $I_{E,C}^{u_{i^*},j^*+1}(i^*,j^*) = I_{E,C}^p(i^*,j^*)$, increase u_{i^*,j^*} by 1.

7: Generate $I_{E,V}(i,j^*)$ using Equation (4.9).

8: Evaluate $I_{CMI}(j)$ for all $j = 1, \cdots, N_p$ using Equation (4.10).

9: **if** $I_{CMI}(j) == 1$ for all $j = 1, \cdots, N_p$

10:   Terminate.

11: **else**

12:   Increase s by 1, position = 1.

13: **end if**

---

the schedule arrangement is completed when the $I_{CMI}(j)$ value for each j converges to one, another reasonable criterion can be possibly used to maximize the increase in the CMI, where the CMI increase can be generated according to the prediction of the CMI. The prediction of the CMI which corresponds to the C2V messages transmitted from type-i check nodes to the neighboring type-j variable nodes can be obtained when all check nodes neighboring v_j, except c_i, provide a fixed C2V mutual information value. The predicted CMI is a function of $I_{E,C}^p(i,j)$ according to Equation (4.10), and can be denoted as $I_{CMI}^p(j, I_{E,C}^p(i,j))$. Therefore, the CMI increase associated with the type-(i,j) edges can be calculated as $r_{CMI}(i,j) = \left(I_{CMI}^p(j, I_{E,C}^p(i,j)) - I_{CMI}(j)\right)$. The criterion of maximizing $r_{CMI}(i,j)$ can also be adopted in the M^2I^2-based algorithm. However, it is found that following the resultant decoding schedule, the average number of updates of the C2V messages varies with the associated variable node degree, and cannot achieve a faster convergence than that based on Equation (4.6). Consequently, a criterion using Equation (4.6) is adopted in the proposed M^2I^2-based algorithm.

The proposed M^2I^2-based algorithm is utilized to arrange the decoding schedule for a code ensemble with the assistance of the extrinsic information transfer chart (EXIT) analysis, which is adopted to evaluate the mutual information based on the assumptions of infinite code length and the statistically independent Gaussian-distributed decoding messages which are exchanged within a cycle-free bipartite graph. Due to these assumptions, the arranged schedule may not be able to provide the fastest convergence speed for the practical finite-length codes with cycles. Consequently, the proposed M^2I^2-based algorithm cannot yet provide the optimum decoding schedule for a practically implemented LDPC code.

4.3.7 High-order Prediction for the M^2I^2-based Algorithm

In Equation (4.5), the difference in mutual information between two adjacent decoding stages, which can be regarded as the order-1 increment, is generated. However, when considering the protograph-based edge-wise schedule, the updated messages in the current decoding stage will affect all the following stages, rather than only the subsequent stage. Consequently, if the mutual information can be predicted through the messages produced during several successive decoding stages, **S** can be more precisely determined for achieving an even faster convergence.

In the determination of $S(s + 1)$ by using the $\mathrm{M}^2\mathrm{I}^2$-based algorithm, all protograph edges can be the candidates for $S(s + 1)$, and, therefore, all the corresponding depth-1 (i.e., $s + 1$) predictions and depth-1 increments in the mutual information should be evaluated. If the predictions of the mutual information for messages which are going to be produced in the $(s + 2)$-th stage, i.e., the depth-2 predictions, are also available, we may be allowed to generate the depth-2 increment, which is defined as the difference between the mutual information for the $(s+2)$-th stage and that for the $(s+1)$-th stage. Since the updates in the $(s + 1)$-th stage can affect the depth-2 predictions, and hence, the depth-2 increment, the messages related to distinct protograph edges will induce different versions for both the depth-2 predictions and the depth-2 increments. If the messages related to type-(i, j) edges have been updated in the $(s + 1)$-th stage, the depth-2 increments in the $(s + 2)$-th stage can be denoted as $r_2^{(i,j)}(h, k)$, for all edges where $b_{h,k} \neq 0$. It is denoted by the superscript (i, j) that these depth-2 increments are calculated in the $(s + 2)$-th stage following the updates to the edge-(i, j) messages in the $(s + 1)$-th stage. In addition, the maximum value of these depth-2 increments can be denoted as $\Delta(i, j)$. It is notable that the updates in the $(s + 2)$-th stage can be selected from the messages related to all protograph edges, and are not limited to those affected by the messages updated in the $(s + 1)$-th stage. However, since repeated updates to the same messages cannot provide any additional mutual information increase, the depth-2 increment $r_2^{(i,j)}(i, j)$ will be excluded from the comparison when determining $\Delta(i, j)$.

In Figure 4.2, we consider the LDPC codes constructed through the use of base matrix $\mathbf{B}$ given in Equation (4.4). The candidates to be updated in the $(s+1)$-th stage include the C2V messages related to all protograph edges. For the messages related to the type-$(2, 1)$ edges which will possibly be updated in the $(s+1)$-th stage, both $r(2, 1)$ and $\Delta(2, 1)$ are generated. When determining $\Delta(2, 1)$, all depth-2 increments in the $(s+2)$-th stage are compared, except for $r_2^{(2,1)}(2, 1)$. For the messages related to type-$(3, 4)$ edges, $r(3, 4)$ and $\Delta(3, 4)$ can be obtained in a similarly manner. The updates to the C2V messages related to the type-$(2, 1)$ and the type-$(3, 4)$ edges will then renew the values of the type-1 and the type-4 variable nodes, respectively. Since the messages related to the type-$(2, 1)$ and the type-$(3, 4)$ edges contribute to different parts of the decoding messages in the $(s + 1)$-th stage, they will produce different versions of both the depth-2 predictions and the depth-2 increments in the $(s + 2)$-th stage. As a result, $\Delta(2, 1)$ and $\Delta(3, 4)$ are generated based on the different versions of the predictions, and their values can also be different.

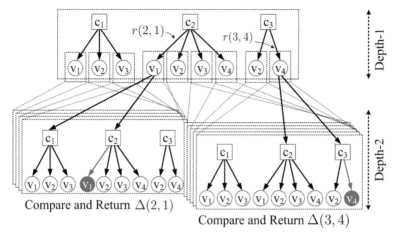

Figure 4.2   The order-2 M^2I^2-based algorithm for the protograph corresponding to the base matrix **B** generated in Equation (4.4).

Moreover, it is possible that $r(2, 1) > r(3, 4)$ but $\Delta(2, 1) < \Delta(3, 4)$, which implies that the updates to the messages related to the type-(2,1) edges can provide a higher level of increase in mutual information in the next stage, i.e., the $(s + 1)$-th stage. However, the updates to the type-$(3, 4)$ messages in the $(s + 1)$-th stage can induce a higher level of increase in mutual information in the stage after the next one, i.e., the $(s + 2)$-th stage. Therefore, the selection for the updates based on the depth-1 increment generated according to Equation (4.5) is suitable for the next decoding stage, but may not be the best strategy for the later decoding stages, which are beyond the scope of the depth-1 increment.

In order to improve the decision result, the order-1 increment and the depth-2 increment can be combined as the order-2 increment, which is given by

$$r_\Delta(i, j) = I^p_{E,C}(i, j) - I^{u_{ij}}_{E,C}(i, j) + \Delta(i, j) = r(i, j) + \Delta(i, j). \quad (4.11)$$

If the order-2 increment $r_\Delta(2, 1)$ achieves the maximum value, the updates to the messages related to the type-$(2, 1)$ edges in the $(s + 1)$-th stage can be guaranteed to provide the maximum increase in mutual information for the later two stages, i.e., the $(s+1)$-th and $(s+2)$-th stages. In fact, a higher order increment can be recursively generated in the form of this order-2 increment, which is presented in detail in Algorithm 2:. If we replace the original order-1 increment $r(i, j)$ used in line 3 of Algorithm 2 with the order-P increment

---

Algorithm 2: The order-P increment generator.

1: **FUNCTION R_GEN**$(\delta, (i,j), (I_{E,C}^{p}(\mu,\lambda), I_{E,C}^{u_{\mu,\lambda}}(\mu,\lambda), I_{E,V}(\mu,\lambda), \forall b_{\mu,\lambda} \neq 0))$

2: Set $r(i,j) = I_{E,C}^{p}(i,j) - I_{E,C}^{u_{i,j}}(i,j)$ and $\Delta(i,j) = 0$.

3: **if** $\delta < P$

4:   Increase δ by 1 and set $I_{E,C}^{u_{i,j}}(i,j) = I_{E,C}^{p}(i,j)$.

5:   Generate $I_{E,V}(a,j)$ using Equation (4.9) for all $c_a \in \mathcal{N}(v_j) \setminus c_i$.

6:   Generate $I_{E,C}^{p}(a,b)$ using Equation (4.8) for all $c_a \in \mathcal{N}(v_j) \setminus c_i$ and $v_b \in \mathcal{N}(c_a) \setminus v_j$.

7:   **for**$(h = 1 \ldots M_p, k = 1 \ldots N_p$, for all $b_{h,k} \neq 0$ and $(h,k) \neq (i,j))$

8:     Generate $r_{\delta}^{(i,j)}(h,k) = $ **R_GEN**$(\delta, (h,k), (I_{E,C}^{p}(\mu,\lambda), I_{E,C}^{u_{\mu,\lambda}}(\mu,\lambda), I_{E,V}(\mu,\lambda),$
        $\forall b_{\mu,\lambda} \neq 0))$.

9:   **end for**

10:   Find $\Delta(i,j) = \max \left\{ r_{\delta}^{(i,j)}(h,k) \mid h = 1, \cdots, M_p, k = 1, \cdots, N_p, (h,k) \neq (i,j) \right\}$.

11: **end if**

12: **RETURN**$(r_{\Delta}(i,j) = r(i,j) + \Delta(i,j))$.

$r_{\Delta}(i,j)$, an order-P M^2I^2-based algorithm can be used, where the increase in the mutual information provided by the updates from the $(s + 1)$-th to the $(s + P)$-th decoding stages can be calculated in order to assist in the determination of $S(s + 1)$.

In Algorithm 2:, an order-P increment is recursively calculated order by order using the subroutine **R_GEN**. The depth index δ is initially set to 1. The edge index (i,j) of the protograph is used to indicate that the subroutine is executed to refine the increment associated with type-(i,j) edges. The mutual information $I_{E,C}^{p}(\mu,\lambda)$, $I_{E,C}^{u_{\mu,\lambda}}(\mu,\lambda)$, and $I_{E,V}(\mu,\lambda)$ for all non-zero elements of the base matrix $\mathbf{B}$, i.e., $b_{\mu,\lambda} \neq 0$, are inherited from the previous state of the recursion, and their values can only be passed to the next state, but cannot affect the values in the previous state.

When $\delta < P$, the value of the refined increment $r_{\Delta}(i,j)$ is evaluated under the assumption that the C2V messages associated with type-(i,j) edges have been updated. To evaluate the next-order increments in line 4, the depth index δ increases by 1, and the predicted C2V mutual information $I_{E,C}^{p}(i,j)$ acts as the mutual information provided by the actual updated messages, i.e., $I_{E,C}^{u_{i,j}}(i,j) = I_{E,C}^{p}(i,j)$. Then, the V2C mutual information transfered from type-j variable nodes to the neighboring check nodes, and the predicted C2V mutual information associated with these neighboring check nodes can be calculated in lines 5 and 6 based on this assumption. Therefore, despite only part of the predicted C2V mutual information being renewed from line 4 to line 6, the increments for all types of edges must be generated, except for any repeated edges. The depth-δ increment $r_{\delta}^{(i,j)}(h,k)$ can be generated for all protograph edges by recursively executing the subroutine **R_GEN** in line 8. The superscript (i, j) denotes that these increments are generated based on the assumption that $I_{E,C}^p(i, j)$ will be provided by the actual updates. The subroutine **R_GEN** will be recursively executed until $\delta = P$. Then, the value of the maximum increment $\Delta(i, j)$ can be determined after the comparison in line 10, and can be used to refine the increment returned to the previous recursive state in line 12.

In order to complete the arrangement of the schedule using the order-P M^2I^2-based algorithm, a depth-P tree search should be performed. Since there are E_B branches on each node, the complexity of the predictions and comparisons is $\mathcal{O}(E_B^P)$. In the search tree, the branches on each node can be considered as the candidates of a list, thus the concept of list decoding [26] can be introduced. It is obvious that there is a trade-off between complexity and performance using different list sizes. Since the M^2I^2-based algorithm is executed offline, and we are more interested in the performance of the resultant schedules, a full list of E_B edges is considered in this work.

4.3.8 Performance Evaluation

As previously mentioned, the M^2I^2-based algorithm can be initiated using any SNR_s values larger than the decoding threshold, and distinct schedules can be obtained for different decoding SNR values. For the rate-0.5 LDPC code used in the G.hn standard, the decoding threshold of the code ensemble is 0.82 dB, then a total five schedules can be arranged, respectively, using SNR_s = 0.82, 0.85, 0.9, 1.4, and 3.0 dB. For the practical length-8640 G.hn code, a meaningful error rate can be achieved when the SNR (E_b/N_0) value is greater than 1.25 dB. Figure 4.3 shows the error rate versus the number of iterations when different schedules are used, where the considered decoding SNR value is 1.4 dB. It can be observed that the decoding processes following different schedules can provide very similar performances, even when the schedule arranged using SNR_s = 3.0 dB is used. Since the decoding performance is not sensitive to the selection of schedules, only a single schedule is used in the decoding for the entire SNR region.

The convergence trajectories for the average CMI I_{CMI} are plotted in Figure 4.4, where the results are respectively obtained using increments from order 1 to order 4. Figure 4.4 also shows the trajectory of the flooding schedule, which is obtained using the original PEXIT (protograph-based extrinsic information transfer) chart analysis [24]. For all update strategies, it can be observed that the protograph-based edge-wise schedule designed using the

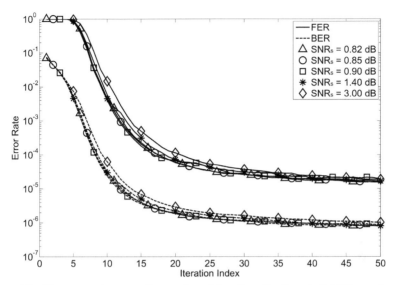

Figure 4.3 The error rate versus the number of iterations for the rate-0.5 G.hn code using different schedules. ($P = 4$, AWGN channel, $E_b/N_0 = 1.4$ dB)

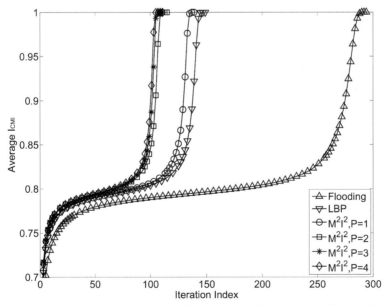

Figure 4.4 The convergence behavior of the average I_{CMI} using different decoding schedules. (AWGN channel and $E_b/N_0 = 0.85$ dB)

M^2I^2-based algorithm based on the order-1 increment requires only half the number of iterations compared to the conventional flooding schedule. With higher order predictions, the convergence speed can be further improved. For this rate-0.5 G.hn code, there is no significant difference in performance when either order-3 or order-4 increments are used, and the number of required iterations is about 40% that of the conventional flooding schedule.

An evaluation of the frame-error rate (FER) and bit-error rate (BER) performances for the schedules arranged using the M^2I^2-based algorithm based on different increment orders is presented in Figure 4.5. It can be observed from Figure 4.4 and Figure 4.5 that the relative improvements in the convergence speed for both the mutual information and the error rate provided by the M^2I^2-based algorithm are consistent, which implies that the M^2I^2-based algorithm is a useful method for arranging the decoding sequence.

The error-rate results for the length-2304, rate-3/4 WiMax code and the length-8640, rate-1/2 G.hn code are shown in Figures 4.6–4.7 and Figure 4.8, respectively. It can be seen from Figure 4.6 that, for the high-rate WiMAX code, the M^2I^2-based algorithm can provide the same FER performances at the same SNR using 25% fewer iterations compared to the number required

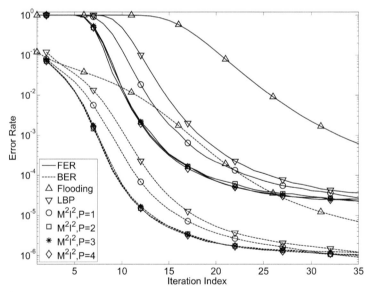

Figure 4.5 The convergence speed for the rate-1/2 length-8640 G.hn [22] code using different decoding schedules. (AWGN channel and $E_b/N_0 = 1.40$ dB)

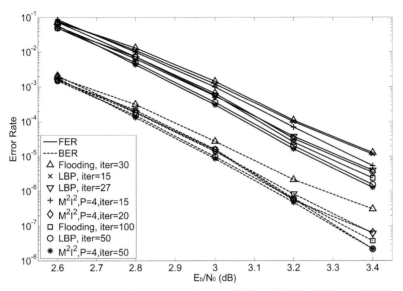

Figure 4.6 The error rate performance for the rate-3/4 length-2304 WiMax code [21] using different decoding schedules.

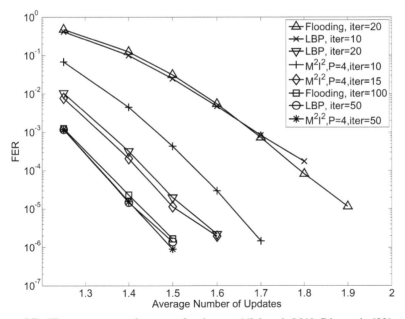

Figure 4.7 The error rate performance for the rate-1/2 length-8640 G.hn code [22] using different decoding schedules.

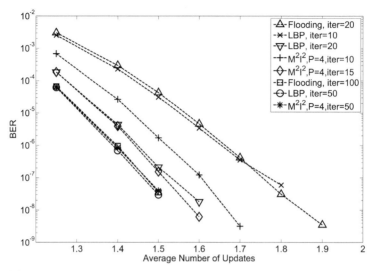

Figure 4.8   The error rate performance for the rate-1/2 length-8640 G.hn code [22] using different decoding schedules.

for the LBP schedule. The FER and BER performance are presented in Figures 4.6 and 4.7, respectively. For the low-rate G.hn code, using the schedule arranged by the M^2I^2-based algorithm consistently reduces the number of required iterations by 25% compared to that required for the LBP schedule, and gains more than 0.2 dB in SNR at an FER of 2×10^{-4} for the same number of iterations.

The reason why the protograph-based edge-wise schedule arranged by using the M^2I^2-based algorithm can provide a faster convergence compared to LBP is described as follows. When LBP is used, the decoding process is scheduled to sequentially follow the order of the types of check nodes in the protograph, and the messages associated with the same row of matrix **B** are updated simultaneously. Therefore, the protograph-based edge-wise schedule can provide more freedom in the schedule design, and hence, a faster convergence in error-rate performance is observed.

4.4 A Reduction of the Complexity for Scheduling Arrangement

It can be observed that, in order to complete the arrangement of the schedule using the order-P M^2I^2-based algorithm, a depth-P tree search should be performed. Since for each decoding stage, the number of the branches on each node is equal to the sizes of the protograph E_B, the complexity of the predictions and comparisons is $\mathcal{O}(E_B^P)$. Then the M^2I^2-based algorithm with high-order prediction can be unattractive due to the exponentially increasing complexity.

In order to reduce the computation requirement of each decoding stage, the concept of list decoding [26] can be introduced. For the search tree, the branches on each node can be considered as the candidates of a list, then for the original M^2I^2-based algorithm, the size of the list can be considered to be the number of edges in the protograph, i.e., E_B. The major difference between the proposed LM^2I^2-based algorithm and the M^2I^2-based algorithm is that only E_L edges are remaining in the list, the complexity of the predictions for the order-P increment can be reduced from $\mathcal{O}(E_B^P)$ to $\mathcal{O}((E_L)^P)$.

For the order-P LM^2I^2-based algorithm, the size of the lists of all the decoding stages are fixed to E_L. For each decoding stage, the E_L edges with maximum increment in C2V mutual information are selected from all the E_B edges as the candidates in the list. The detailed process of the proposed LM^2I^2-based algorithm is presented as Algorithm 3:. It is worth noting that, when $E_L = E_B$, the operations of the LM^2I^2-based algorithm are identical to those of the original M^2I^2-based algorithm, then the M^2I^2-based algorithm can be regarded as a special case of the LM^2I^2-based algorithm.

Algorithm 3: Proposed LM^2I^2-based schedule arrangement.

Initialization:
  Set the size of the list equals E_L
  Set the prediction order equals P.
1: **for**$(i = 1, \cdots, M_p, j = 1, \cdots, N_p$, and $b_{i,j} \neq 0)$
2:   Generate $I_{E,C}^p(i,j)$.
3:   Generate $r(i,j) = I_{E,C}^p(i,j) - I_{E,C}(i,j)$.
4:   **for** the E_L edges with the maximum $r(i,j)$:
5:     Recursively evaluate $r_\Delta(i,j)$ using Equation (4.11);
6:   **end for**
7: **end for**
8: Determine (i^*, j^*) based on Equation (4.6), $S(s) = (i^*, j^*)$.
9: $I_{E,C}(i^*, j^*) = I_{E,C}^p(i^*, j^*)$.
10: Evaluate $I_{CMI}(j)$ for all $j = 1, \cdots, N_p$.
11: **if** $I_{CMI}(j) == 1$ for all $j = 1, \cdots, N_p$
12:   Terminate.
13: **else**
14:   Increase s by 1, position = 1.
15: **end if**

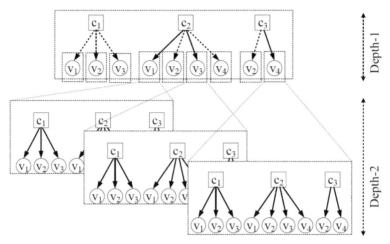

Figure 4.9   The search tree for $E_L = 3$.

For the base matrix **B** in Equation (4.4), Figure 4.9 shows the example of that the proposed LM^2I^2-based algorithm, where $E_L = 3$ and $P = 2$. The list size is 3, assuming that $r(2, 1)$, $r(2, 3)$ and $r(3, 4)$ are the three maximum increments in the C2V mutual information of the first decoding stage. Then only three branches expand in the second decoding stage. Comparing to the case shown in Figure 4.2, where the original M^2I^2-based algorithm was applied, the required number of predictions has been obviously reduced.

The benefit obtained from the additional parameter E_L is not limited to reduced the complexity. When the available computational resource is limited, or the same amount of complexity is allowed, the prediction order can be increased when a small list size determines the complexity of each decoding stage. Then, with large P and small E_L the proposed LM^2I^2-based algorithm can be used to arrange a protograph-based edge-wise schedule that provides faster decoding convergence than using the M^2I^2-based algorithm.

4.4.1 Performance Evaluation for the LM^2I^2-based Algorithm

In order to evaluate the influence of the list size on the decoding performance, distinct schedules were arranged using the LM^2I^2-based algorithm based on different E_L values. When the LM^2I^2-based algorithm is applied to the protograph of the length-8640 G.hn code [22], where the number of the edges is $E_B = 76$, the trajectories of the I_{CMI} values evaluated using these schedules are shown in Figure 4.10. It can be observed that with a small

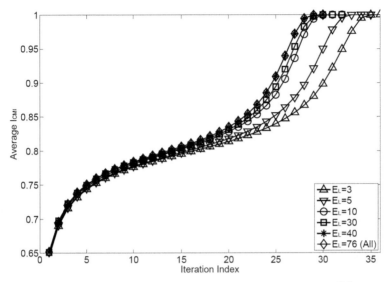

Figure 4.10 The convergence behavior of the average I_{CMI} when the LM^2I^2-based algorithm using different list sizes of E_L is adopted. (rate-0.5 G.hn code, $P = 4$, $E_B = 76$, additive white Gaussian noise (AWGN) channel, $E_b/N_0 = 0.85$ dB).

list size, such as $E_L = 3$ or $E_L = 5$, the convergence speed for the I_{CMI} values slows down significantly. As the size of the list increases, e.g., when $E_L > 10$, the convergence speed for the I_{CMI} values is close to that obtained using the full-size list, i.e., using the original M^2I^2-based algorithm. The schedule arranged using $E_L = 40$ can provide a performance that is the same as that arranged using $E_L = 76$; however, it reduces the required number of predictions to $1/10$.

The FER (and BER) results versus the number of iterations are shown in Figure 4.11, where the decoding process follows a schedule using different E_L values. A trend similar to that presented using the I_{CMI} values can be observed in Figure 4.11, where the error-rate performance of the LBP schedule is also appended. It is worth noting that a much faster convergence can be achieved, even the decoding schedules are arranged using the low-cost LM^2I^2-based algorithm.

The example provided in Figuers. 4.10 and 4.11 demonstrates that, considering about $1/10$ of the computation complexity, a decoding schedule can be arranged using the proposed LM^2I^2-based algorithm that has the same performance as that arranged using the original M^2I^2-based algorithm.

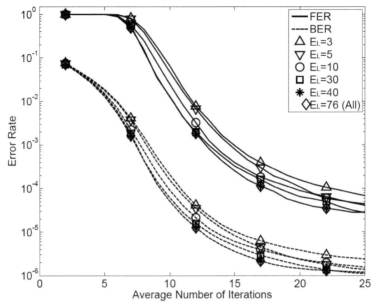

Figure 4.11   The convergence behavior for the frame-error rate (FER) and bit-error rate (BER) performance when schedules arranged using different list sizes of E_L. (rate-0.5 G.hn code, $P = 4$, $E_B = 76$, AWGN channel, $E_b/N_0 = 1.40$ dB).

Consequently, if the same computation complexity is adopted in the LM^2I^2-based algorithm, a much higher prediction order is allowed to enhanced the performance of the resultant schedules.

For the (2640, 1320) Margulis code [27], the number of the edges on its protograph is $E_B = 720$. If the decoding schedule is arranged using the original M^2I^2-based algorithm, only a small prediction order P can be adopted, due to the high computational complexity. In contrast, a small list size E_L can be selected when the proposed LM^2I^2-based algorithm takes over, and a higher prediction order is possibly adopted.

Figure 4.12 shows the average I_{CMI} convergence behavior for the schedules arranged using the LM^2I^2-based algorithm when different combinations of the list size E_L and the prediction order P are adopted. Consider the schedule arranged using all edges, i.e., $E_L = 720$, with a prediction order $P = 2$, a total branch number of $720^2 = 518400$ is expanded as the candidates for determining the indicator of each decoding stage. If the prediction order is increased to $P = 3$ and the list size is reduced to $E_L = 90$, a faster convergence can be achieved, but the total number of candidates is

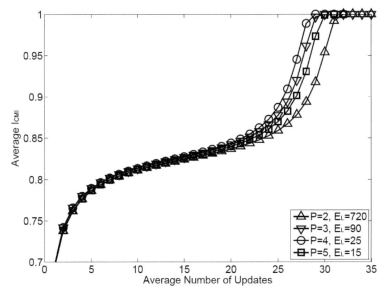

Figure 4.12 The convergence behavior of the average I_{CMI} when the LM^2I^2-based algorithm using different list sizes of E_L and different prediction orders of P is adopted. (rate-0.5 Margulis code, $E_B = 720$, AWGN channel, $E_b/N_0 = 1.12$ dB)

also increased to $90^3 = 729000$. The decoding convergence can be further accelerated when the combination of $P = 4$ and $E_L = 25$ is adopted, along with a reduction in the number of candidates as $25^4 = 390625$. However, the convergence cannot be guaranteed to further speed up as the prediction order increased. For the combination of $P = 5$ and $E_L = 15$, the convergence speed is even slower than that of the combination of $P = 3$ and $E_L = 90$.

Figure 4.13 shows the FER (and BER) performances versus the number of iterations, which corresponds to the I_{CMI} convergence behavior shown in Figure 4.12. Again, a trend similar to that presented using the I_{CMI} values can be observed. For all the combination of P and E_L values, the resultant decoding schedules can achieve a much faster convergence speed compared to that of the LBP schedule.

The examples considering the G.hn standard code and the Margulis code show that, with a proper list size of E_L, the LM^2I^2-based algorithm can be used to arrange a schedule which provides almost the same decoding performance as that provided by the schedule arranged using the original M^2I^2-based algorithm, and the computation complexity can be significantly reduced. On the other hand, when the computational resource is limited, a

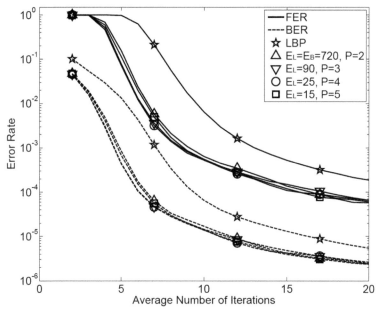

Figure 4.13   The convergence behavior for the FER and BER performance when schedules arranged using different list sizes and prediction orders. (rate-0.5 Margulis code, AWGN channel, $E_b/N_0 = 2.1$ dB)

faster convergence decoding can be achieved using the schedule arranged through the proposed LM^2I^2-based algorithm with a proper combination of P and E_L values.

4.5 Lower Error Floor via Schedule Diversity

For LDPC codes, when the error floor appears in the high-SNR region, the majority of erroneously decoded frames are due to trapping sets. An (a, b) trapping set can be triggered when its a variable nodes are simultaneously in error and its b check nodes connecting to these variable nodes are not satisfied. Figure 4.14 shows an example of a Tanner graph, where the variable nodes $\{v_3, v_4, v_5, v_6\}$ and the check node c_2 potentially construct a $(4, 1)$ trapping set. Trapping sets with a small a value can be more easily triggered. If it also contains a few unsatisfied check nodes, i.e., those with a small b value, which can generate contributing decoding messages, then it is difficult to solve the erroneously decoded bits in this trapping set. Therefore, the trapping sets with

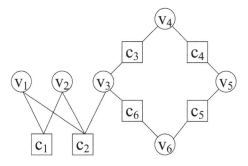

$S_T = \{(2,3),(3,4),(4,5),(5,2),(5,5),(4,4),(3,3),(2,2),(1,2),(0,0),(0,1),(1,0),(1,1),...\}$

$S_C = \{(2,3),(3,4),(4,5),(5,2),(0,0),(0,1),(1,0),(1,1),(1,2),(5,5),(4,4),(3,3),(2,2),...\}$

	LLR Values of $\{V_1, V_2, V_3, V_4, V_5, V_6\}$	Hard Decision Results
Initial	$\{1.07, \ 1.01, \ -0.3, \ -0.4, \ -0.3, \ -0.1\}$	$\{0,0,1,1,1,1\}$
Using S_T, 40 iterations	$\{-0.14, 0.23, -49.87, -49.17, -49.17, -49.17\}$	$\{1,0,1,1,1,1\}$
Using S_C, 1 iteration	$\{1.060, 1.060, 0.305, 0.002, 0.002, 0.002\}$	$\{0,0,0,0,0,0\}$

Figure 4.14 An example of a Tanner graph, where $\{v_3, v_4, v_5, v_6\}$ and $\{c_2\}$ form a $(4, 1)$ trapping set.

small (a, b) values are most harmful to the error floor performance and, hence, are called *dominant trapping sets*.

When the sub-optimum iterative BP algorithm is used in LDPC decoding, a single received frame can converge at a variety of speeds using different schedules, such as flooding, LBP, or the protograph-based edge-wise schedule. This implies that the values of the decoding messages for the same received frame are certainly different when distinct schedules are used. According to the hard decision results from the received LLR values, a $(4, 1)$ trapping set is triggered. Using the decoding schedule that follows the order of sequence $\mathbf{S}_T$, the errors are unable to be corrected after 40 iterations, and, even worse, the number of erroneously decoded bits is increased, as shown in Figure 4.14. In contrast, if decoding sequence $\mathbf{S}_C$ is used, all erroneously decoded bits are corrected within a single iteration.

4.5.1 Decoding Scheme Combined with Schedule Diversity

The example demonstrated in Figure 4.14 shows that the schedules indeed affect the decoding results. It can be expected that, when multiple decoding schedules are available for a single received frame, schedule diversity can be achieved. In the high SNR region, a relatively low error rate can be achieved using a single decoding schedule for a well-constructed code. This implies that trapping sets rarely occur, even when only a single decoding schedule is available. If multiple decoding schedules can be applied to a single received frame, an even lower probability of trapping sets can be achieved.

In order to increase the probability of successful decoding, multiple decoding attempts, respectively using distinct schedules, are applied to a single received frame. The channel values are loaded in the initial state of each decoding attempt. The decoding messages are then updated following a predetermined schedule, e.g., S, arranged using the M^2I^2-based algorithm. If the received codeword cannot be successfully decoded, the decoding messages are reset, and the channel values are reloaded as the next decoding attempt starts. In order to achieve a diverse decoding result, a schedule different from S must be used. The distinct schedule S_{R1} for this new decoding attempt can be obtained by randomly permutating the elements in S. The new schedule can be regarded as a permutated, or an interleaved, version of S. The random permutations can be generated using a pseudo-random scrambler, which is commonly used in code division multiple access (CDMA) systems [28]. Pseudo-random scrambling codes can provide low-correlation sequences. Following the low-correlation sequences S and S_{R1}, there is a high probability that diverse decoding results can be obtained during these two decoding attempts.

Figure 4.15 shows the error rate performances achieved when the proposed schedule diversity is applied to the (2640, 1320) Margulis code [27]. If only a single decoding attempt is used, both FER and BER floor can be observed when the SNR value is larger than 2.3 dB. As the number of decoding attempts increases, the error-rate floors can be significantly lowered. It can also be observed that the improvement in the FER is more obvious than for the BER. The reason is that, on average, a single erroneously decoded frame contains more erroneously decoded bits as the number of decoding attempts increases, as shown in Figure 4.16. For the Margulis code, the dominant trapping sets are $(12, 4)$ and $(14, 4)$, which account for about 75% and 23% of the error floor performance, respectively [14]. Hence, it is expected that the average number of erroneously decoded bits in each erroneously decoded frame following the first decoding attempt $(T = 1)$ will be about $12 \times 75\% + 14 \times 23\% = 12.22$ in the error-floor region. The result observed from Figure 10 at SNR = 2.5 dB and $T = 1$ is consistent with this expectation. It can be deduced that the majority of the erroneously decoded frames are likely due to the dominant trapping sets. Because of the diverse decoding results, there is a low probability that the dominant

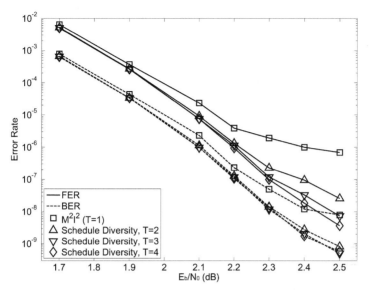

Figure 4.15 The error rate performance for the rate-1/2 length-2640 Margulis code [27] when a different number of decoding attempts are applied, where the number of decoding attempts is denoted as T. The maximum number of iterations is limited to 50 for each decoding attempt.

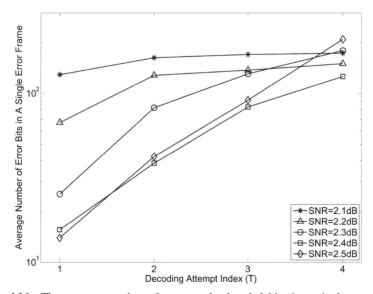

Figure 4.16 The average number of erroneously decoded bits in a single erroneously decoded frame.

trapping sets are triggered at exactly the same point in the decoding process. Consequently, the erroneously decoded frames are rarely a result of the dominant trapping sets following multiple decoding attempts. On the other hand, there is a higher probability that the decoding fails due to the high-weight error patterns, e.g., other trapping sets with larger a values, or even other legal codewords. Therefore, it can be observed from Figure 4.16 that in the high-SNR region, the average number of erroneously decoded bits in a single erroneously decoded frame continuously increases as more decoding attempts are executed. At the same time, an improvement in FER of more than two orders in magnitude can be achieved compared to the first decoding attempt, as shown in Figure 4.15. As a result, the proposed schedule diversity indeed mitigates the effect of the dominant trapping sets, and is able to lower the error floor.

4.5.2 Comparison with Other Error Floor Lowering Techniques

In addition to the proposed schedule diversity, the decoding methods proposed in [29] are able to lower the error floor with the assistance of the scheduling technique, where both LBP and ANS [29] (Approximate Node-wise Scheduling) are employed. Although the switching of schedules is introduced in the LBP/ANS decoders proposed in [29], the decoding messages are not cleared and the channel values are not reloaded. Since a complete decoding attempt includes the loading of the channel values, there is only a single decoding attempt applied to each received codeword for the LBP/ANS decoders. Moreover, the capability of lowering the error floor is derived from the dynamic characteristics of the ANS, rather than the schedule switching.

As with the proposed schedule diversity, the clear and reload operations have also been used in the backtracking algorithm presented in [15] to reset the decoder, when the previous decoding attempt failed. However, in contrast to the proposed schedule diversity, the sign of the channel value corresponding to one of the suspicious error bits is flipped before the new decoding attempt starts. The new decoding attempt is then initiated following the same LBP schedule. In other words, for the backtracking algorithm, the channel values used in each decoding attempt are not exactly the same as the original values, and only one kind of schedule is used. The differences between these three decoding methods are shown in Figure 4.17.

Figure 4.18 shows the FER performances when different decoding schemes are applied to the (2640, 1320) Margulis code, where the maximum

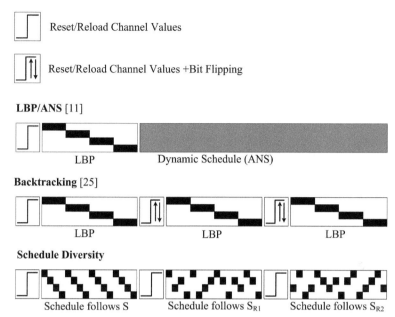

Figure 4.17 Comparison of different error-floor lowering techniques.

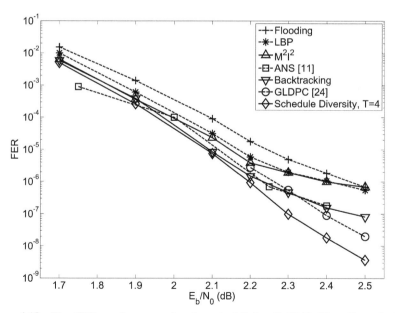

Figure 4.18 The FER performance for the rate-1/2 length-2640 Margulis code [27] using different decoding schemes. For the backtracking algorithm, the number of decoding attempts is 6.

number of iterations is limited to 50. Since the LBP/ANS decoders in [29] have not been applied to the Margulis code, and the capability of lowering the error floor comes from the ANS decoder, only the results for the ANS decoder are included in Figure 4.18. The error floor can be observed when the SNR is higher than 2.3 dB for the decoding schedules based on Flooding, LBP, and the schedule arranged using the proposed M^2I^2-based algorithm. Although the ANS decoder [29] and the backtracking algorithm are able to significantly lower the error floor, the proposed schedule diversity can provide an even lower error floor. It is worth noting that the proposed schedule diversity can provide a better FER performance than that of generalized LDPC (GLDPC) decoding [14], which is usually regarded as a benchmark for error-floor-lowering techniques when applied to the Margulis code.

4.6 Remarks

In this chapter, both the convergence speed and the error-floor performance are improved using the fixed decoding schedules techniques. The maximum mutual information increase (M^2I^2)-based algorithm based on the idea of providing the maximum increase in the mutual information is proposed to arrange the schedules for faster convergence.

On the other hand, the *schedule diversity* can be achieved using multiple decoding attempts with multiple decoding schedules, and the error floor can be lowered without the knowledge of the trapping sets. In addition, adopting the schedule arranged using the M^2I^2-based algorithm in the first decoding attempt, and combining it with schedule diversity for the following decoding attempts, both the convergence speed in the waterfall region and the error rate in the error-floor region can be improved.

References

[1] Gallager, R. G. (1963). *Low-Density Parity-Check Codes*. Cambridge, MA: MIT Press.

[2] 3GPP (2016). *Draft Report of 3GPP TSG RAN WG1 #87 v0.1.0*. Reno, NV: A Global Initiative.

[3] Zhang, J., and Fossorier, M. (2005). Shuffled iterative decoding. *IEEE Trans. Commun.* 53, 209–213.

[4] Yeo, E., Pakzad, P., Nikolic′, B., and Anantharam, V. (2001). "High throughput low-density parity-check decoder architectures," in *Proceedings of the 2001 Global Conference on Communications*, San Antonio, TX, 3019–3024.

[5] Mansour, M. M., and Shanbhag, N. R. (2003). High-throughput LDPC decoders. *IEEE Trans. Very Large Scale Integr. Syst.* 11, 976–996.

[6] Hocevar, D. (2004). "A reduced complexity decoder architechture via layered decoding of LDPC codes," in *Proceedings of the Signal Process Systems (SIPS)*, Lorient, 107–112.

[7] Ueng, Y.-L., Yang, C.-J., Wang, K.-C., and Chen, C.-J. (2010). A multi-mode shuffled iterative decoder architecture for high-rate RS-LDPC codes. *IEEE Trans. Circuits Syst. I Reg. Pap.* 57, 2790–2803.

[8] Wang, Y.-L., Ueng, Y.-L., Peng, C.-L., and Yang, C.-J. (2011). Processing-task arrangement for a low-complexity full-mode WiMAX LDPC Codec. *IEEE Trans. Circuits Syst. I Reg. Pap.* 58, 415–428.

[9] Casado, A. V., Griot, M., and Wesel, R. (2007). "Informed dynamic scheduling for belief-propagation decoding of LDPC codes," in *Proceedings of the IEEE International Conference Communication (ICC)*, Glasgow.

[10] Wang, Y.-L., Ueng, Y.-L., Peng, C.-L., and Yang, C.-J. (2011). Processing-task arrangement for a low-complexity full-mode WiMAX LDPC Codec. *IEEE Trans. Circuits Syst. I* 58, 415–428.

[11] Ueng, Y.-L., Yang, B.-J., Yang, C.-J., Lee, H.-C., and Yang, J.-D. (2013). An efficient multi-standard LDPC decoder design using hardware-friendly shuffled decoding. *IEEE Trans. Circuits Syst. I* 60, 743–756.

[12] Lee, H.-C., and Ueng, Y.-L., (2014). LDPC decoding scheduling for faster convergence and lower error floor. *IEEE Trans. Commun.* 62, 3104–3113.

[13] MacKay, D., and Postol, M. (2003). Weaknesses of Margulis and Ramanujan-Margulis low-density parity-check codes. *Electron. Notes Theor. Comput. Sci.* 74, 97–104.

[14] Han, Y., and Ryan, W. (2009). Low-floor decoders for LDPC codes. *IEEE Trans. Commun.* 57, 1663–1673.

[15] Kang, J., Huang, Q., Lin, S., and Abdel-Ghaffar, K. (2011). An iterative decoding algorithm with backtracking to lower the error floors of LDPC codes. *IEEE Trans. Commun.* 59, 64–73.

[16] Chen, X. W., Yu, H., and Xu, Y. Y. (2009). "Optimized decoding schedule for irregular LDPC codes," in *Proceedings of the 2009 IET International Communication Conference on Wireless Mobile and Computing (CCWMC)*, Shanghai, 261–264.

[17] Kim, M. K., Kim, D. H., and Lee, Y. H. (2012). "Serial scheduling algorithm of LDPC decoding for multimedia transmission," in *Proceedings*

of the 2012 IEEE International Symposium on Broadband Multimedia Systems and Broadcasting (BMSB), London, 1–4.

[18] Sharon, E., Litsyn, S., and Goldberger, J. (2007). Efficient serial message-passing scheduleds for LDPC decoding. *IEEE Trans. Inf. Theory* 53, 4076–4091.

[19] Sharon, E., Presman, N., and Litsyn, S. (2009). Convergence analysis of generalized serial message-passing schedules. *IEEE J. Sel. Areas Commun.* 27, 1013–1024.

[20] IEEE 802.11 (). *Wireless LANs WWiSE Proposal: High throughput Extension to the 802.11 Standard*. Rome: *IEEE*.

[21] IEEE (2005). *IEEE 802.16e WiMAX Standard*. Rome: IEEE.

[22] *ITU-TG.hn Standard forWiredHome Networking* (2017). Available at: http://www.homegridforum.org/home/

[23] Thorpe, J. (2003). *Low-Density Parity-Check (LDPC) Codes Constructed from Protographs*. Pasadena, CA: JPL, 42–154.

[24] Liva, G., and Chiani, M. (2007). "Protograph LDPC codes design based on EXIT analysis," in *Proceedings of the 2007 IEEE GlobeCom Conference,* Washington, DC, 3250–3254.

[25] ten Brink, S., Kramer, G., and Ashikhmin, A. (2004). Design of low-density parity-check codes for modulation and detection. *IEEE Trans. Commun.* 52, 670–678.

[26] Guruswami, V. (2001). *List Decoding of Error-correcting Codes*. Ph.D. dissertation, MIT, Cambridge, MA.

[27] Margulis, G. A. (1982). Explicit constructions of graphs without short cycles and low-density codes. *Combinatorica* 1, 71–78.

[28] Ahmed, I. (2004). *Scrambling Code Generation for WCDMA on the StarCore SC140/SC1400 Cores*. San Jose, CA: Altera.

[29] Casado, A. V., Griot, M., and Wesel, R. (2010). LDPC decoders with informed dynamic scheduling. *IEEE Trans. Commun.* 58, 3470–3479.

5

Location Template Matching on Rigid Surfaces for Human–Computer Touch Interface Applications

Nguyen Q. Hanh, V. G. Reju and Andy W. H. Khong

School of Electrical and Electronic Engineering, Nanyang Technological University, Singapore

Abstract

In this chapter we discuss a category of signal processing techniques called location template matching (LTM), which is widely used to convert daily objects such as glass panels and tabletops into human–computer touch interfaces. In such applications, a single sensor is mounted on a rigid surface to capture vibrations induced by users' impacts. LTM methods compare location-dependent features of the vibration signal generated by a user with a pre-collected library to estimate the impact location. In the most basic form of LTM, the signal itself is utilized as a location-dependent feature where each library signal is correlated with the received signal to determine the best match. More advanced techniques utilize the mechanical model of vibration in plate to extract other useful location-dependent features. This chapter reviews different LTM techniques arising from the time-reversal theory and the classical model for flexural vibration on thin plates. Specifically, we focus our discussion on which features are utilized for location matching and the respective matching measures. The algorithms will also be compared against each other to highlight their advantages and disadvantages.

5.1 Introduction

Advancements in technology have been revolutionizing the way humans interact with computers. Traditional input devices such as keyboard and mouse are being replaced by new human–computer interfaces (HCIs) which are more human-centric. Among such HCIs, touch interfaces make up a significant portion of the market due to the advent of mobile devices such as smart-phones and tablets. In addition, touch interfaces nowadays are not limited to pre-designed products but extended to objects that are readily available in daily life [1] such as tabletops and glass panels. This is where main stream touch technologies such as capacitive- and resistive-sensing see theirs limitations due to the high manufacturing cost, which exponentially increases with the size of the device. Research in recent years has therefore been focusing on alternative solutions for flexible extension of touch interfaces to large surfaces [2–6]. This chapter focuses on such an alternative that is based on vibration technology. Using low-cost sensor(s) to capture surface vibrations, impacts generated by fingers or styluses on the surfaces can be localized. This offers a cost-effective and flexible means of converting large surfaces into touch interfaces. Figure 5.1(a) illustrates such a conversion of an existing glass table-top into a human–computer touch interface by mounting low-cost Murata PKS1-4A10 piezoelectric shock sensors. Such a sensor is shown in Figure 5.1(b). Vibration signals captured by the sensors, due to user interaction on the surface, are forwarded to a computer for estimating the interacting point using an appropriate signal processing algorithm.

Time-difference-of-arrival (TDOA) and location template matching (LTM) are among the most popular approaches to address the problem of source localization on rigid surfaces [2, 7]. In the former approach, the sensor outputs are analyzed to estimate the TDOA between each sensor pair, which, in turn, determines a hyperbolic curve with the foci being the sensor locations. Since each such curve passes through the impact location, the location is estimated as the intersection point of all the curves. On rigid surfaces, conventional methods for TDOA estimation such as the generalized cross-correlation [8] suffer severely from velocity dispersion and multipath. Pre-processing is required to reduce the effect of dispersion, where signal components corresponding to a specific narrow frequency band are isolated [9]. The TDOA between a sensor pair is subsequently obtained as the difference in time-of-arrivals (TOAs) of the respective components [6, 10]. By focusing on the arrival of each signal component, any distortion due to multipath is also mitigated. Localization using this approach, however, often requires a low-noise environment since the accuracy of the algorithm deteriorates significantly when the arrival portion of the signal is submerged in the background noise.

The LTM approach, as opposed to TDOA-based localization, utilizes signal distortions due to dispersion and multipath to determine the impact

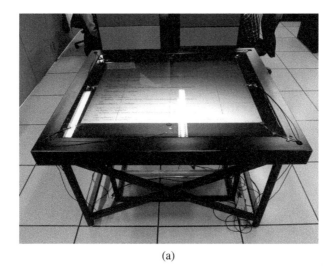

(a)

(b)

Figure 5.1 (a) A setup for human–computer interface application where a transparent glass plate is placed on top of a flat-panel display. Sensors are mounted along the edges of the plate to pick up vibration signal and forward to a computer for processing. (b) A low-cost Murata PKS1-4A10 piezoelectric shock sensor used in the setup.

location. Since such distortions are, in theory, unique to the source-sensor propagation path, the distorted signal received by the sensor possesses features related to the impact location. These features can therefore be utilized to localize the impact [11, 12]. LTM localization involves two phases: calibration and test. In the first phase, impacts are exerted at a set of pre-defined calibration points. A single shock sensor is mounted on the surface to capture vibrations induced by the impacts. Location-dependent features are subsequently extracted from the sensor-received signals and stored in a library. During testing, a received (test) signal would have the same features extracted and compared with those stored in the library. The impact location is subsequently estimated by finding the closest match. One of the most popular approaches for LTM is based on the time-reversal theory, where the signal itself is utilized as a location-dependent feature. The best match is then determined from correlating each library signal with the test signal. More advanced LTM techniques utilize the mechanical model of vibration in thin plates to extract other useful location-dependent features. For example, the infinite impulse response (IIR)-LTM method models the signal using the all-pole IIR response filter and employs the filter coefficients as a location-dependent feature [11]. The Zak transform-based LTM (Z-LTM) algorithm, on the other hand, tracks the amplitude variation of the signal in the time-frequency domain using the Zak transform and quantifies such variation for utilization in location matching [12]. Such features, however, are sensitive to noise. As a result, high signal-to-noise ratio (SNR) signals are required for accurate localization of the impact point. The band-limited component (BLC)-LTM algorithm is the most recent attempt, which combines both the time-reversal and classical-plate theories to achieve robust performance in the presence of band-limited noise [13].

This chapter is organized as follows: The general procedure for determination of an impact location using the LTM approach is discussed in the section "LTM for Impact Localization on Solids." The section "Time-reversal Theory-based LTM" reviews the use of the time-reversal theory in LTM, while the section "Classical Plate Theory-based LTM" discusses the emerging adoption of the classical plate theory in LTM on rigid surfaces. The BLC-LTM algorithm is discussed in the section "Noise-robust LTM Using Band-limited Components" and concluding remarks are given in the section "Concluding Remarks."

5.2 LTM for Impact Localization on Solids

Impact localization via LTM involves two stages, calibration and test, as illustrated in Figure 5.2. In what follows, the superscript $\mathcal{L}_q$, $q = 1, \ldots, Q$, is used to denote the q-th library data/location to be defined during the calibration process, where Q is the total number of calibration locations on the rigid surface. These locations are denoted by $(u^{\mathcal{L}_1}, v^{\mathcal{L}_1}), \ldots, (u^{\mathcal{L}_Q}, v^{\mathcal{L}_Q})$. In a typical LTM set-up, the determined set of calibration points forms a grid on the surface, as depicted in Figure 5.3. During calibration, vibrations are generated at each of the grid points using either a finger or a mechanical tapper. The vibration signals $x^{\mathcal{L}_q}(n)$, $1 \leq n \leq N$, captured by a single sensor mounted at location (u_o, v_o), are stored in a library $\{x^{\mathcal{L}_q}(n); q = 1, \ldots, Q\}$. During the test phase, assuming an impact is generated at one of the calibrated

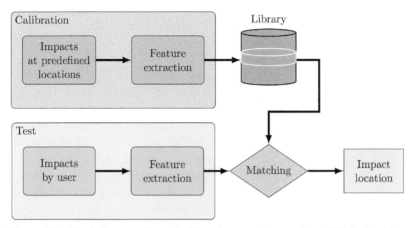

Figure 5.2   Block diagram of a typical location template matching (LTM) algorithm.

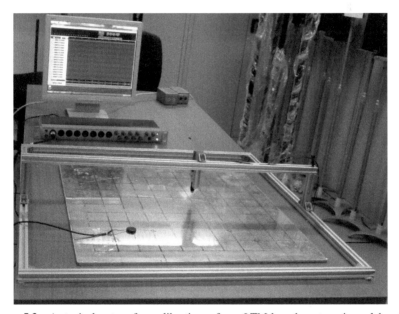

Figure 5.3   A typical setup for calibration of an LTM-based system in a laboratory environment.

locations, $(u_\mathrm{s}, v_\mathrm{s}) \equiv (u^{\mathcal{L}_{\widetilde{q}}}, v^{\mathcal{L}_{\widetilde{q}}})$ for some $1 \le \widetilde{q} \le Q$, the objective of an LTM algorithm is to determine $(u_\mathrm{s}, v_\mathrm{s})$ given the induced vibration signal $x(n)$.

Based on the assumption that an impact-induced signal preserves information about its source location, LTM-based algorithms focus on the extraction of location-specific information from the sensor-received signal. Such information encompasses the signature that distinguishes impacts at one location from those made at other locations. In the test phase, the signature of each signal stored in the library $\{x^{\mathcal{L}_q}(n); q = 1, \ldots, Q\}$ is compared with that of $x(n)$. The library signal with the signature most similar to that of $x(n)$ is considered the *matching* signal. The source location of $x(n)$ is subsequently determined as that of the matching library signal. The success of an LTM algorithm in impact localization depends on which location-dependent features are extracted and how they are used in the matching process. In the following sections, LTM algorithms and their considerations of which feature to exploit as the signature for the signal will be reviewed.

5.3 Time-reversal Theory-based LTM

The time-reversal theory, introduced in [14], allows one to convert a *divergent* wave emanating from an acoustic source into a *convergent* wave focusing on the same source. Application of this theory in impact localization on solid plates has been extensively studied in the literature [7, 15–19]. As will be shown later in this section, time-reversal-based impact localization involves correlating a received test signal with those pre-collected and stored in a library. This effectively utilizes the signal itself as a location-dependent feature. Due to the need for computing the cross-correlation between the test and library signals, time-reversal-based LTM is also known as cross-correlation-based LTM (CC-LTM).

With reference to the time-reversal theory, an impact generated by a user at an arbitrary point (u, v) is assumed to be sufficiently brief so that it can be approximated by the Dirac delta function, i.e., $s(n) \approx \delta(n)$. It follows that during calibration, where an impact is generated at each location $(u^{\mathcal{L}_q}, v^{\mathcal{L}_q})$, the signal $x^{\mathcal{L}_q}(n)$ received at the sensor is given by

$$x^{\mathcal{L}_q}(n) \approx \delta(n) * h^{\mathcal{L}_q}(n)$$
$$= h^{\mathcal{L}_q}(n), \tag{5.1}$$

where $*$ is the convolution operator and $h^{\mathcal{L}_q}(n)$ is the impulse response between the impact location $(u^{\mathcal{L}_q}, v^{\mathcal{L}_q})$ and the sensor location (u_o, v_o). It is worth noting that $h^{\mathcal{L}_q}(n)$ is determined by the propagation path from $(u^{\mathcal{L}_q}, v^{\mathcal{L}_q})$ to (u_o, v_o) and is therefore unique for each $(u^{\mathcal{L}_q}, v^{\mathcal{L}_q})$.

During the test phase, an impact generated at one of the calibrated locations $(u_s, v_s) \equiv (u^{\mathcal{L}_{\tilde{q}}}, v^{\mathcal{L}_{\tilde{q}}})$ for some location index $1 \leq \tilde{q} \leq Q$ yields a sensor output $x(n)$, which approximates $h^{\mathcal{L}_{\tilde{q}}}(n)$. According to the time-reversal theory, if $x(n)$ is time-reversed and "emitted" from the sensor location (u_o, v_o), the vibration signal at each calibrated location $(u^{\mathcal{L}_q}, v^{\mathcal{L}_q})$ will be given by

$$\Psi_{x,h^{\mathcal{L}_q}}(n) = x(-n) * h^{\mathcal{L}_q}(n)$$
$$\approx h^{\mathcal{L}_{\tilde{q}}}(-n) * h^{\mathcal{L}_q}(n). \tag{5.2}$$

This, in fact, corresponds to the cross-correlation between $h^{\mathcal{L}_{\tilde{q}}}(n)$ and $h^{\mathcal{L}_q}(n)$, and can be rewritten as

$$\Psi_{x,h^{\mathcal{L}_q}}(n) = \sum_{\ell=1}^{N} h^{\mathcal{L}_{\tilde{q}}}(n) h^{\mathcal{L}_q}(n + \ell). \tag{5.3}$$

It can easily be seen that $\Psi_{x,h^{\mathcal{L}_q}}(n)$ achieves its maximum if and only if $h^{\mathcal{L}_{\tilde{q}}}(n) = h^{\mathcal{L}_q}(n)$, which is equivalent to $(u^{\mathcal{L}_{\tilde{q}}}, v^{\mathcal{L}_{\tilde{q}}}) \equiv (u^{\mathcal{L}_q}, v^{\mathcal{L}_q})$. Therefore, the impact location can be determined as

$$\tilde{q} = \text{argmax}_{1 \leq q \leq Q} \{\max \Psi_{x,h^{\mathcal{L}_q}}(n)\}. \tag{5.4}$$

Recall from Equation (5.1) that $h^{\mathcal{L}_q}(n)$ has been approximated by $x^{\mathcal{L}_q}(n)$. The function $\Psi_{x,h^{\mathcal{L}_q}}(n)$ is therefore equivalent to $\Psi_{x,x^{\mathcal{L}_q}}(n)$, which quantifies the similarity/dissimilarity between $x(n)$ and $x^{\mathcal{L}_q}(n)$. Other similarity measures can also be utilized in place of $\Psi_{x,x^{\mathcal{L}_q}}(n)$. One such measure is

$$\tilde{\Psi}_{x,x^{\mathcal{L}_q}}(n) = \frac{1}{\sigma_x \sigma_{x^{\mathcal{L}_q}}} \Psi_{x,x^{\mathcal{L}_q}}(n), \tag{5.5}$$

where the standard deviations of $x(n)$ and $x^{\mathcal{L}_q}(n)$, denoted by σ_x and $\sigma_{x^{\mathcal{L}_q}}$, respectively, are introduced to compensate for the variation in signal energy across different user interactions [2]. Other methods to enhance the reliability of the similarity measure are discussed in [18].

5.4 Classical Plate Theory-based LTM

While the time-reversal theory can be utilized in various applications, for the specific case of impact localization on a rigid plate, the performance can be improved by incorporating signal propagation models during algorithm development. A survey of theories on wave propagation on solid plates can be found in [20]. Within the scope of this chapter, however, only models pertaining to thin rectangular solid plate are of concern. Such models include the classical plate theory [21,22], the Mindlin theory (which includes corrections for both rotary inertia and shear effects) [23], and the Rayleigh–Lamb theory [24, 25]. Among these models, the classical plate theory is the simplest and can provide a sufficiently good approximation of the vertical displacement of the plate surface due to an impact as long as the vibration is contained within a sufficiently low-frequency band [20]. As reported in [12] and [16], for touch-interface applications, components of a vibration caused by a finger or a metal stylus are of frequencies less than 10 kHz, and the classical plate theory can therefore be utilized.

In what follows, we will review the classical plate theory for flexural vibrations on a rectangular solid plate of dimension $L_x \times L_y \times L_z$. Assume that an impact is exerted at a location (u_s, v_s) on the plate surface. This will induce a vibration signal $x(u_s, v_s; u_o, v_o, t)$ at the sensor location (u_o, v_o). Employing Kirchoff's assumptions for linear, elastic small-deflection theory of thin plates, the governing differential equation for $x(u_s, v_s; u_o, v_o, t)$ is given by [21]

$$
\dot{D}\nabla^4 x(u_s, v_s; u_o, v_o, t) + \mu \frac{dx(u_s, v_s; u_o, v_o, t)}{dt}
$$
$$
+ \rho L_z \frac{d^2 x(u_s, v_s; u_o, v_o, t)}{dt^2} = p(u_s, v_s; u_o, v_o, t), \quad (5.6)
$$

where $p(u_s, v_s; u_o, v_o, t)$ is the lateral load at (u_o, v_o), ρ the material density, μ the absorption coefficient, $\dot{D} = EL_z^3/12(1-\nu^2)$ the plate's stiffness, E the Young's modulus, ν the Poisson's ratio, and the biharmonic operator $\nabla^4 = (\partial^4/\partial u^4) + 2(\partial^4/\partial u^2 \partial v^2) + (\partial^4/\partial v^4)$.

By modeling the impact exerted at (u_s, v_s) as an impulse, the lateral load at (u_o, v_o) is given as

$$
p(u_s, v_s; u_o, v_o, t) = \alpha \delta(t) \delta(u_o - u_s) \delta(v_o - v_s), \quad (5.7)
$$

where $\delta(\cdot)$ is the Dirac delta function, and α denotes the strength of the impact. The solution to Equation (5.6) can be obtained for a simply supported plate with zero initial condition as [11]

$$
x(u_s, v_s; u_o, v_o, t) = \frac{4\alpha}{\rho L_x L_y L_z} \sum_{l=1}^{\infty} \sum_{b=1}^{\infty} \frac{\Lambda_{lb}(u_s, v_s; u_o, v_o) e^{-0.5\tilde{\mu}t} \sin(\varpi_{lb} t)}{\varpi_{lb}},
$$
$$
(5.8)
$$

where the vibration at (u, v) is expressed as an infinite sum of vibration modes (l, b) with $l, b \in \mathbb{Z}^+$ being the mode indices. Each mode of vibration is determined by three parameters: the shape function

$$\Lambda_{lb}(u_s, v_s; u_o, v_o) = \sin(\beta_l u_s) \sin(\beta_b v_s) \sin(\beta_l u_o) \sin(\beta_b v_o), \qquad (5.9)$$

the reduced absorption coefficient $\widetilde{\mu} = \mu/(\rho L_z)$, and the mode frequency $\varpi_{lb} = \sqrt{\omega_{lb}^2 - (\widetilde{\mu}/2)^2}$, where

$$\omega_{lb} = (\beta_l^2 + \beta_b^2)\sqrt{\frac{D}{\rho L_z}}, \qquad (5.10)$$

and $\beta_l = \pi l/L_x$, $\beta_b = \pi b/L_y$. Equation (5.9) implies that the sensor location imposes different scaling effect on the mode amplitude via $\Lambda_{lb}(u_s, v_s; u_o, v_o)$ corresponding to each mode (l, b). In addition, the scaling effect is also dependent on the impact location (u_s, v_s). Figure 5.4 depicts the deterministic shape of Λ_{lb} as a function of (u_s, v_s) where its maximum amplitude is normalized to 1.

In what follows, LTM techniques which utilize the model in Equation (5.8) will be reviewed. By analyzing this model, different location-dependent

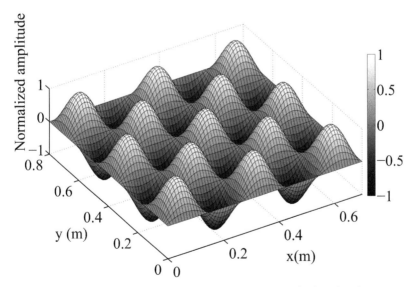

Figure 5.4   The shaping function Λ_{lb} for the mode of vibration $(l, b) = (5, 5)$ [after [12]].

features can be extracted from the signal for use in location matching. Details on how such features are derived will be discussed, in addition to the measures deployed in these LTM methods that quantify the similarity among extracted features. One common factor about these techniques is that theoretical derivations are performed on the continuous-time version of the signals while actual extractions and measure computations are performed on the digitized signals.

5.4.1 All-pole Filter Model-based LTM (AP-LTM)

The impulse response of the vibration sensor can be described by a second-order linear differential equation. By integrating this sensor model to the mechanical model of flexural vibration in thin plates, it has been shown in [11] that the sensor output is a function of frequency. In addition, its spectrum can be utilized in matching $x(n)$ to the library $x^{\mathcal{L}_q}(n)$ via determination of their dominant frequencies. Dominant-frequency extraction from the spectrum of a signal is achieved by using the all-pole model for IIR filters to model the signal. Denoting the all-pole coefficients for $x(n)$ by α_p, $p = 1, \ldots, \mathcal{N}$, where $\mathcal{N}$ is the prediction order, model fitting is achieved by minimizing the difference between $x(n)$ and its estimate $\widehat{x}(n) = \sum_{p=1}^{\mathcal{N}} \alpha_p x(n-p)$, given by

$$\mathcal{J} = \sum_{n=1}^{N} (\varepsilon(n))^2, \tag{5.11}$$

where $\varepsilon(n) = x(n) - \widehat{x}(n)$. The spectrum of $x(n)$ can now be estimated as

$$|\mathcal{P}_x(e^{\jmath\omega})|^2 = \frac{\sigma_\varepsilon^2}{|1 + \alpha_1 e^{-\jmath\omega} + \cdots + \alpha_{\mathcal{N}} e^{-\jmath\omega}|^2}, \tag{5.12}$$

where σ_ε^2 is the variance of $\varepsilon(n)$. Factorizing the denominator in Equation (5.12), the spectrum is given as

$$|\mathcal{P}_x(e^{\jmath\omega})|^2 = \frac{\sigma_\varepsilon^2}{|(e^{\jmath\omega} - e^{\jmath\omega_1}) \ldots (e^{\jmath\omega} - e^{\jmath\omega_{\mathcal{N}}})|^2}, \tag{5.13}$$

where the poles $\omega_1, \ldots, \omega_{\mathcal{N}}$ correspond to dominant frequencies in the spectrum of $x(n)$. Estimation of dominant frequencies via all-pole modeling for two typical vibration signals are illustrated in Figure 5.5. Here, IIR filter of order $\mathcal{N} = 3$ is utilized in modeling. The filter coefficients α_p are evaluated for signals received from impacts generated at locations (0.41, 0.20) and

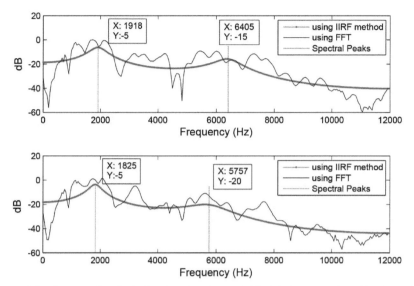

Figure 5.5 Spectra of typical vibration signals overlaid with the respective all-pole model approximations. Here the test signals are generated at locations (0.41 m, 0.20 m) and (0.42 m, 0.20 m) while the sensor is mounted at (0.05 m, 0.25 m). The all-pole IIR filter is of order $\mathcal{N} = 3$ [after [11]].

(0.42, 0.20), with the sensor mounted at (0.05, 0.25) (all coordinates are in meters). As can be seen, each spectrum estimate (in red) peaks at the dominant frequencies of the corresponding signal. In addition, it is useful to note that even though the impact locations are spatially close, their dominant frequencies differ from each other.

Since the dominant frequencies are unique for each signal, they can be utilized as its signature for location matching. Equivalently, the coefficients can also be employed as the signal's signature. Denoting the coefficients for $x^{\mathcal{L}_q}$ by $\alpha_p^{\mathcal{L}_q}$, $p = 1, \ldots, \mathcal{N}$, the dissimilarity measure between $x(n)$ and $x^{\mathcal{L}_q}(n)$ is given by

$$\mathcal{D}_{x,x^{\mathcal{L}_q}}^{\text{AP}} = \sum_{p=1}^{\mathcal{N}} \left(\alpha_p - \alpha_p^{\mathcal{L}_q} \right)^2. \tag{5.14}$$

The source location of $x(n)$ is subsequently estimated as $(u^{\mathcal{L}_{\tilde{q}}}, v^{\mathcal{L}_{\tilde{q}}})$ that yields the minimum dissimilarity, as given by

$$\tilde{q} = \operatorname{argmin}_q \mathcal{D}_{x,x^{\mathcal{L}_q}}^{\text{AP}}. \tag{5.15}$$

5.4.2 Zak Transform for Time-frequency-based LTM (Z-LTM)

According to the classical plate theory, the library signals $x^{\mathcal{L}_q}(n)$ can be expressed using Equation (5.8) as

$$x^{\mathcal{L}_q}(t) = \frac{4}{\rho L_x L_y L_z} \sum_{l=1}^{\infty} \sum_{b=1}^{\infty} \frac{\Lambda_{lb}(u^{\mathcal{L}_q}, v^{\mathcal{L}_q})e^{-0.5\tilde{\mu}t}\sin\left(\varpi_{lb}t\right)}{\varpi_{lb}}, \qquad (5.16)$$

where

$$\Lambda_{lb}(u^{\mathcal{L}_q}, v^{\mathcal{L}_q}) = \sin(\beta_l u^{\mathcal{L}_q})\sin(\beta_b v^{\mathcal{L}_q})\sin(\beta_l u_o)\sin(\beta_b v_o). \qquad (5.17)$$

Note that, since only one sensor at (u_o, v_o) is utilized, the dependency of Λ_{lb} and $x^{\mathcal{L}_q}(t)$ on (u_o, v_o) is implied without being explicitly shown in Equation (5.16). In addition, $\Lambda_{lb}(u^{\mathcal{L}_q}, v^{\mathcal{L}_q})/\varpi_{lb}$ serves as the amplitude of the vibration mode (l, b). Since Λ_{lb} is unique to the impact location $(u^{\mathcal{L}_q}, v^{\mathcal{L}_q})$, the variance across time for each mode is also dependent on $(u^{\mathcal{L}_q}, v^{\mathcal{L}_q})$. By defining

$$\vartheta_{lb}(t) = \frac{\Lambda_{lb}(u^{\mathcal{L}_q}, v^{\mathcal{L}_q})e^{-0.5\tilde{\mu}t}\sin(\varpi_{lb}t)}{\varpi_{lb}}, \qquad (5.18)$$

the variance across time for each mode of vibration (l, b) is given by [12]

$$\begin{aligned}
\sigma_{lb}^2(u^{\mathcal{L}_q}, v^{\mathcal{L}_q}) &= \mathrm{E}_{\vartheta_{lb}}\left\{t^2\right\} - [\mathrm{E}_{\vartheta_{lb}}\left\{t\right\}]^2 \\
&= \int_0^{\infty} t^2 \vartheta_{lb}(t)\mathrm{d}t - \left[\int_0^{\infty} t\vartheta_{lb}(t)\mathrm{d}t\right]^2 \\
&= \frac{\Lambda_{lb}(u^{\mathcal{L}_q}, v^{\mathcal{L}_q})}{\varpi_{lb}}\left(\frac{\tilde{\mu}^2\varpi_{lb}^2(3 - \varpi_{lb} - \tilde{\mu}) - \varpi_{lb}^5 + 0.375\tilde{\mu}^4}{\omega_{lb}^2}\right),
\end{aligned}$$
$$(5.19)$$

where $\mathrm{E}_{\vartheta_{lb}}\{.\}$ is the expectation operator with density function $\vartheta_{lb}(t)$. Being a function of $(u^{\mathcal{L}_q}, v^{\mathcal{L}_q})$, the variance σ_{lb}^2 can be employed as a location-dependent feature of the signal.

To analyze the variance across time for each mode frequency, it is desirable to utilize a time-frequency representation (TFR) that can track the instantaneous change in time and frequency content of the signal. More specifically, the choice of TFR should allow one to obtain the instantaneous energy by summing its energy distribution across all frequencies, and the energy density spectrum by summing across time. These properties are referred to as the time- and frequency-marginal conditions [26]. Here the discrete Zak transform (DZT) [27, 28] has been chosen to represent the signal in the time-frequency domain as this transform satisfies the marginal conditions. The DZT of a received signal $x(n)$ is obtained by first shaping the length-N vector $\mathbf{x} = [x(1), \ldots, x(N)]^{\mathrm{T}}$ into a $L \times M$ matrix defined as

$$\boldsymbol{\mathcal{X}}_{L \times M} = \begin{bmatrix} x(1) & x(2) & \cdots & x(M) \\ x(M+1) & x(M+2) & \cdots & x(2M) \\ \vdots & \vdots & \ddots & \vdots \\ x((L-1)M+1) & x((L-1)M+2) & \cdots & x(LM) \end{bmatrix},$$

(5.20)

where $LM = N$. The DFT is subsequently applied to each column of $\boldsymbol{\mathcal{X}}_{L \times M}$, resulting in a DZT matrix $\boldsymbol{\mathcal{Z}}_x = [\mathcal{Z}_x(k, m)]$ of size $K \times M$ with K being the number of frequency bins. Each element of the DZT matrix is given by [29]

$$\mathcal{Z}_x(k, m) = \sum_{\ell=1}^{L} x(m + \ell M)e^{-j\ell\omega_k},$$

(5.21)

where $\omega_k = 2\pi k/K$ is a normalized angular frequency, $1 \leq k \leq K$. The variance of $\boldsymbol{\mathcal{Z}}_x$ across time for each frequency ω_k is then estimated as

$$\widehat{\sigma}_x^2(k) = \frac{1}{M} \sum_{m=1}^{M} \left[|\mathcal{Z}_x(k, m)| - \frac{1}{M} \sum_{m'=1}^{M} |\mathcal{Z}_x(k, m')| \right]^2,$$

(5.22)

and the vector

$$\widehat{\boldsymbol{\sigma}}_x^2 = \left[\widehat{\sigma}_x^2(1), \ldots, \widehat{\sigma}_x^2 \left(\frac{K}{2} \right) \right]^{\mathrm{T}}$$

(5.23)

defines a location-dependent feature for impacts made at location $(u^{\mathcal{L}_q}, v^{\mathcal{L}_q})$. Note that due to the symmetric property of the DFT, it suffices to compute $\widehat{\sigma}_x^2(k)$ only up to $k = K/2$. A (dis)similarity measure between the test signal $x(n)$ and a library signal $x^{\mathcal{L}_q}(n)$ is subsequently given as

$$\mathcal{D}_{x, x^{\mathcal{L}_q}}^{\mathrm{Zak}} = \sum_{k=1}^{K/2} \left(\widehat{\sigma}_x^2(k) - \widehat{\sigma}_{x^{\mathcal{L}_q}}^2(k) \right)^2,$$

(5.24)

and the index of the location where the test signal is generated is obtained as

$$\widetilde{q} = \mathrm{argmin}_{1 \leq q \leq Q} \mathcal{D}_{x, x^{\mathcal{L}_q}}^{\mathrm{Zak}}.$$

(5.25)

5.5 Noise-robust LTM Using Band-limited Components

The time-reversal theory and the classical plate theory have been utilized extensively in existing LTM algorithms. In the approach based on the time-reversal theory, the signal itself is deployed as the signature for matching. On the other hand, integration of the classical plate theory for vibration modeling helps to determine certain location-dependent features that can be utilized in location matching. The use of such features not only reduces computational complexity but also enhances performance compared to time-reversal-based algorithms, where the whole signal is utilized during the signature-matching process. This advantage is, however, only achieved when the SNR is sufficiently high as features extracted from the signal are, in general, prone to corruption by the background noise. To this end, time-reversal-based LTM is more robust due to the inclusion of the whole signal in the matching process. The BLC-LTM algorithm, combining the advantages of both approaches, is a noise robust LTM algorithm, and will be discussed in this section.

5.5.1 Band-limited Components as Location-dependent Features

Recall from section "Classical Plate Theory-based LTM" that the vibration induced at (u_o, v_o) due to an impact at (u_s, v_s) is given, according to the classical plate theory, as [11]

$$x(u_s, v_s, t) = \frac{4\alpha}{\rho L_x L_y L_z} \sum_{l=1}^{\infty} \sum_{b=1}^{\infty} \frac{\Lambda_{lb}(u_s, v_s)e^{-0.5\tilde{\mu}t}\sin{(\varpi_{lb}t)}}{\varpi_{lb}}. \qquad (5.26)$$

Note that, compared to Equation (5.8), the dependency of x and Λ_{lb} on (u_o, v_o) is implied instead of being shown explicitly in Equation (5.26) for simplicity in notation. It can be seen from Equation (5.26) that the frequency ϖ_{lb} and amplitude Λ_{lb}/ϖ_{lb} identify each vibration mode. As a result, being the sum of infinitely many modes, $x(u_s, v_s, t)$ can be fully represented by the set of value pairs $\Omega = \{(\Lambda_{lb}, \varpi_{lb}) \mid l, b = 1, 2, \dots\}$. Recall from Equation (5.9), which is reproduced here for convenience,

$$\Lambda_{lb}(u_s, v_s) = \sin(\beta_l u_s)\sin(\beta_b v_s)\sin(\beta_l u_o)\sin(\beta_b v_o), \qquad (5.27)$$

that the amplitude Λ_{lb}/ϖ_{lb} of each mode is unique to (u_s, v_s) and (u_o, v_o). This implies that any non-empty subset of Ω can serve as a location-dependent feature of $x(u_s, v_s, t)$.

Consider a combination of modes with frequencies limited within some band $B_f = (f - \Delta f, f + \Delta f)$, given as

$$z_f(u_{\mathrm{s}}, v_{\mathrm{s}}, t) = \frac{4\alpha}{\rho L_{\mathrm{x}} L_{\mathrm{y}} L_{\mathrm{z}}} \sum_{\varpi_{lb} \in B_f} \frac{\Lambda_{lb}(u_{\mathrm{s}}, v_{\mathrm{s}}) e^{-0.5\tilde{\mu}t} \sin(\varpi_{lb} t)}{\varpi_{lb}}. \quad (5.28)$$

This is a *band-limited component* of $x(u_{\mathrm{s}}, v_{\mathrm{s}}, t)$, and corresponds to a subset of Ω, denoted by $\Omega_f = \{(\Lambda_{lb}, \varpi_{lb}) \mid \varpi_{lb} \in B_f; l, b \in \mathbb{Z}^+\}$. Due to the equivalence of $z_f(u_{\mathrm{s}}, v_{\mathrm{s}}, t)$ and Ω_f, the component can be employed as a signature for the impact location $(u_{\mathrm{s}}, v_{\mathrm{s}})$.

The BLC-LTM algorithm utilizes BLCs as location-dependent features for location matching. In addition, BLC-LTM also employs a measure that utilizes the cross-correlation for evaluating the similarity between any two feature profiles. The measure exploits the dispersion across the BLCs to improve the robustness of the algorithm to background noise.

5.5.2 The BLC-LTM Algorithm

In the context of solid plates, any signal distortion is attributed to the effects of multipath and dispersion. Multipath distortion is due to vibrations reflected at the medium boundary and interfering with the vibration reaching the sensor via the direct path. Such distortion is, therefore, unique to the geometrical relation between the source, sensor and boundary. Distortion due to dispersion, on the other hand, is the result of propagation velocity being a function of vibration frequency; the further a signal travels, the more its frequency components are separated in space, causing the signal to flatten. While CC-LTM exploits the uniqueness in the shape of the signal brought about by these two phenomena, the dispersion property can be further exploited to propose a new matching measure for better localization performance.

Recall from section "Band-limited Components as Location-dependent Features" that BLCs of a signal can serve as its location-dependent features. The process to obtain a matching measure that involves these features is shown in Figure 5.6. Here, the BLCs are first extracted by filtering the signal using K band-pass filters with frequency responses $H_k(f)$ where $k = 1, \ldots, K$ is the band index. Inputs to the band-pass filters correspond to the test signal $\mathbf{x} = [x(1), \ldots, x(N)]^{\mathrm{T}}$ and the library signals $\mathbf{x}^{\mathcal{L}_q} = [x^{\mathcal{L}_q}(1), \ldots, x^{\mathcal{L}_q}(N)]^{\mathrm{T}}$, $q = 1, \ldots, Q$. Outputs of the filters (the BLCs) are respectively denoted by $\mathbf{z}_k$ and $\mathbf{z}_k^{\mathcal{L}_q}$. To evaluate the similarity between

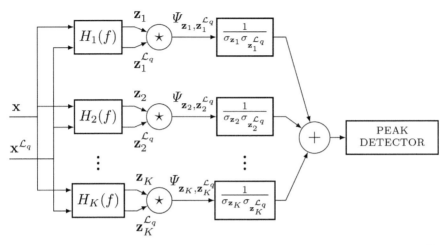

Figure 5.6   The matching measure between the test signal $\mathbf{x}$ and a library signal $\mathbf{x}^{\mathcal{L}_q}$ deployed in the band-limited component (BLC)-LTM algorithm [after [13]].

$\mathbf{x}$ and $\mathbf{x}^{\mathcal{L}_q}$, consider the correlation between the two corresponding BLCs, given by

$$\Psi_{\mathbf{z}_k, \mathbf{z}_k^{\mathcal{L}_q}}(\ell) = \sum_{n=1}^{N} z_k(\ell) z_k^{\mathcal{L}_q}(\ell + n),$$ (5.29)

and denote the lag corresponding to its maximum as

$$\ell^{\mathrm{max}}_{z_k, z_k^{\mathcal{L}_q}} = \mathrm{argmax}_l \, \Psi_{\mathbf{z}_k, \mathbf{z}_k^{\mathcal{L}_q}}(\ell).$$ (5.30)

Utilizing the velocity dispersion property of the material, it is possible to show that the highest peaks of $\Psi_{\mathbf{z}_k, \mathbf{z}_k^{\mathcal{L}_q}}(\ell)$ align at the same lag *for all* band index k only for a certain set of library data. This set of $\mathbf{x}^{\mathcal{L}_q}$ corresponds to locations that are equidistant from the sensor where the distance is that between the sensor and the true impact location. In other words, it can be shown for all pairs $k \neq k'$, $1 \leq k, k' \leq K$, that

$$\begin{cases} \ell^{\mathrm{max}}_{z_k, z_k^{\mathcal{L}_q}} = \ell^{\mathrm{max}}_{z_{k'}, z_{k'}^{\mathcal{L}_q}}, & \text{if } d(u_{\mathrm{s}}, v_{\mathrm{s}}) = d(u^{\mathcal{L}_q}, v^{\mathcal{L}_q}), & (5.31\mathrm{a}) \\ \ell^{\mathrm{max}}_{z_k, z_k^{\mathcal{L}_q}} \neq \ell^{\mathrm{max}}_{z_{k'}, z_{k'}^{\mathcal{L}_q}}, & \text{if } d(u_{\mathrm{s}}, v_{\mathrm{s}}) \neq d(u^{\mathcal{L}_q}, v^{\mathcal{L}_q}), & (5.31\mathrm{b}) \end{cases}$$

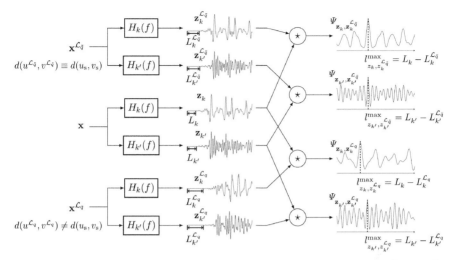

Figure 5.7 Cross-correlations between the test signal $\mathbf{x}$ and the library signals $\mathbf{x}^{\mathcal{L}_{\tilde{q}}}$ and $\mathbf{x}^{\mathcal{L}_{q}}$. For $\mathbf{x}^{\mathcal{L}_{\tilde{q}}}$, which corresponds to the true impact location, the peaks of the correlation functions are aligned whereas for $\mathbf{x}^{\mathcal{L}_{q}}$, which corresponds to a location other than that of the source, the peaks are not aligned [after [13]].

where $d(u', v') = \|(u', v') - (u_o, v_o)\|$ is the distance from an arbitrary point (u', v') to the sensor. This property is subsequently exploited for a signal matching measure which achieves a high value only when the test signal $\mathbf{x}$ is compared against the library data from the same impact location. A detailed proof of Equations (5.31a) and (5.31b) is given below.

Proof. In this method, the bandwidth of the K non-overlapping band-pass filters is chosen sufficiently narrow so that, for each band, the difference in velocity among the frequency components constituting the respective BLC is negligible. The k-th BLC can therefore be considered as propagating with a constant velocity c_k that is determined by the center frequency f_k of the corresponding band-pass filter, i.e., $c_k = c(f_k)$, $k = 1, \ldots, K$. On the other hand, the non-overlapping filters are also selected in such a way that velocity dispersion exist between different BLCs, i.e., $c_k \neq c_{k'}$ if $k \neq k'$, $1 \leq k$, $k' \leq K$.

Consider any two BLCs $\mathbf{z}_k$ and $\mathbf{z}_{k'}$ of $\mathbf{x}$, $k \neq k'$, $1 \leq k, k' \leq K$. When arriving at the sensor via direct path, the BLCs would have propagated a distance $d(u_s, v_s)$ with velocities c_k and $c_{k'}$. The arrival times of $\mathbf{z}_k$ and

$\mathbf{z}_{k'}$ are therefore given by $\tau_k = d(u_s, v_s)/c_k$ and $\tau_{k'} = d(u_s, v_s)/c_{k'}$, respectively. The difference in arrival times is then given by

$$\tau_{k,k'} = \tau_k - \tau_{k'}$$
$$= d(u_s, v_s)\left(\frac{1}{c_k} - \frac{1}{c_{k'}}\right). \tag{5.32}$$

By similar argument, the difference between arrival times for the BLCs $\mathbf{z}_k^{\mathcal{L}_q}$ and $\mathbf{z}_{k'}^{\mathcal{L}_q}$ of $\mathbf{x}^{\mathcal{L}_q}$ is given by

$$\tau_{k,k'}^{\mathcal{L}_q} = d(u^{\mathcal{L}_q}, v^{\mathcal{L}_q})\left(\frac{1}{c_k} - \frac{1}{c_{k'}}\right). \tag{5.33}$$

Note that, as opposed to the conventional TDOA concept, which refers to the difference in arrival times between signals of *two* sensors, here $\tau_{k,k'}$ or $\tau_{k,k'}^{\mathcal{L}_q}$ refers to the difference in arrival times between BLCs of *one* sensor-received signal.

In theory, $\tau_{k,k'}$ can also be obtained as the difference between the silent-period lengths of the data frames $\mathbf{z}_k$ and $\mathbf{z}_{k'}$, and so as $\tau_{k,k'}^{\mathcal{L}_q}$ with respect to $\mathbf{z}_k^{\mathcal{L}_q}$ and $\mathbf{z}_{k'}^{\mathcal{L}_q}$. It follows that

$$L_k - L_{k'} = \tau_{k,k'}$$
$$= d(u_s, v_s)\left(\frac{1}{c_k} - \frac{1}{c_{k'}}\right) \tag{5.34}$$

and

$$L_k^{\mathcal{L}_q} - L_{k'}^{\mathcal{L}_q} = \tau_{k,k'}^{\mathcal{L}_q}$$
$$= d(u^{\mathcal{L}_q}, v^{\mathcal{L}_q})\left(\frac{1}{c_k} - \frac{1}{c_{k'}}\right), \tag{5.35}$$

where L_k, $L_{k'}$, $L_k^{\mathcal{L}_q}$ and $L_{k'}^{\mathcal{L}_q}$ are the lengths of the silent periods of $\mathbf{z}_k$, $\mathbf{z}_{k'}$, $\mathbf{z}_k^{\mathcal{L}_q}$, and $\mathbf{z}_{k'}^{\mathcal{L}_q}$, respectively. Note that even though z_k and $z_{k'}$ are extracted from the same signal, their silent periods L_k and $L_{k'}$ are different due to velocity dispersion. This is also the case for $L_k^{\mathcal{L}_q}$ and $L_{k'}^{\mathcal{L}_q}$.

Equations (5.34) and (5.35) imply that when $\mathbf{x}$ and $\mathbf{x}^{\mathcal{L}_q}$ correspond to locations that are equidistant from the sensor, i.e., $d(u_s, v_s) = d(u^{\mathcal{L}_q}, v^{\mathcal{L}_q})$, the above differences in arrival times are the same, i.e.,

$$L_k - L_{k'} = L_k^{\mathcal{L}_q} - L_{k'}^{\mathcal{L}_q}, \qquad 1 \le k, k' \le K, \tag{5.36}$$

which can be rewritten as

$$L_k - L_k^{\mathcal{L}_q} = L_{k'} - L_{k'}^{\mathcal{L}_q}, \qquad 1 \le k, k' \le K. \tag{5.37}$$

Here it is worth noting that since the multipath effect is neglected and only dispersion has been taken into account, the cross-correlation between $\mathbf{z}_k$ and $\mathbf{z}_k^{\mathcal{L}_{\tilde{q}}}$ achieves its maximum at the lag that compensates for the difference between their arrival times, i.e.,

$$\ell_{\mathbf{z}_k, \mathbf{z}_k^{\mathcal{L}_q}}^{\max} = L_k - L_k^{\mathcal{L}_q}. \tag{5.38}$$

The result in Equation (5.37) therefore implies Equation (5.31a). On the contrary, when $\mathbf{x}^{\mathcal{L}_q}$ is from a location where $d(u_s, v_s) \neq d(u^{\mathcal{L}_q}, v^{\mathcal{L}_q})$, the equalities in Equations (5.36) and (5.37) no longer hold, which leads to Equation (5.31b). The above implication is depicted in Figure 5.7, where $\mathbf{x}^{\mathcal{L}_{\tilde{q}}}$ and $\mathbf{x}$ are from locations that are equidistant from the sensor, whereas $\mathbf{x}^{\mathcal{L}_q}$ is from a location of a different distance from the sensor.

Note that there may exist multiple library data corresponding to locations which lie on the circle centered at the sensor with radius $d(u_s, v_s)$. While the BLCs' correlation peaks are aligned for all such locations, if multipath is taken into account, these peaks are low in amplitude except only for the library data collected at the true impact location. This is due to the uniqueness of reflections from the boundary with respect to impact location. □

It is important to note that Equation (5.31b) does *not* apply for the case of non-dispersive signals. In the absence of velocity dispersion, it follows that $1/c_k - 1/c_{k'} = 0, 1 \le k, k' \le K$. As a result, Equations (5.34) and (5.35) always imply $L_k - L_{k'} = 0$ and $L_k^{\mathcal{L}_q} - L_{k'}^{\mathcal{L}_q} = 0$. These can be rewritten as $L_k - L_k^{\mathcal{L}_q} = L_{k'} - L_{k'}^{\mathcal{L}_q}$, or equivalently

$$\ell_{\mathbf{z}_k, \mathbf{z}_k^{\mathcal{L}_q}}^{\max} = \ell_{\mathbf{z}_{k'}, \mathbf{z}_{k'}^{\mathcal{L}_q}}^{\max}, \qquad q = 1, \ldots, Q. \tag{5.39}$$

This implies that the correlations $\Psi_{\mathbf{z}_k, \mathbf{z}_k^{\mathcal{L}_q}}(\ell)$, $k = 1, \ldots, K$, always have their maximum peaks aligned at the same lag, irrespective of the library data vector used for comparison.

With the above, a measure for signal matching is given as

$$\Upsilon_q(\ell) = \sum_{k=1}^{K} \frac{1}{\sigma_{\mathbf{z}_k} \sigma_{\mathbf{z}_k^{\mathcal{L}_q}}} \Psi_{\mathbf{z}_k, \mathbf{z}_k^{\mathcal{L}_q}}(\ell), \tag{5.40}$$

where $\sigma_{\mathbf{z}_k}$ and $\sigma_{\mathbf{z}_k^{\mathcal{L}q}}$ are the standard deviations of $\mathbf{z}_k$ and $\mathbf{z}_k^{\mathcal{L}q}$, respectively. This measure makes use of the fact that the maxima of $\Psi_{\mathbf{z}_k, \mathbf{z}_k^{\mathcal{L}q}}(\ell)$, $k = 1, \dots, K$, are aligned at the same lag when $(u_{\mathrm{s}}, v_{\mathrm{s}}) = (u^{\mathcal{L}q}, v^{\mathcal{L}q})$, resulting in a dominant maximum of $\Upsilon_q(\ell)$. On the other hand, when $(u_{\mathrm{s}}, v_{\mathrm{s}}) \neq (u^{\mathcal{L}q}, v^{\mathcal{L}q})$, the maxima of $\Psi_{\mathbf{z}_k, \mathbf{z}_k^{\mathcal{L}q}}$ are smaller and misaligned. As a result, $\Upsilon_q(\ell)$ exhibits single/multiple peak(s) with magnitude(s) significantly lower than that of the dominant peak when there is a match. The impact location is therefore identified as $(u_{\mathrm{s}}, v_{\mathrm{s}}) = (u^{\mathcal{L}\tilde{q}}, v^{\mathcal{L}\tilde{q}})$, where

$$\tilde{q} = \mathrm{argmax}_{1 \leq q \leq Q}\{\max \Upsilon_q(\ell)\}. \tag{5.41}$$

This correlation-based measure is depicted in Figure 5.6.

In what follows, the behavior of $\Psi_{\mathbf{z}_k, \mathbf{z}_k^{\mathcal{L}q}}(\ell)$ in the presence of noise is investigated. For simplicity, assume that calibration is conducted in a noise-free environment while noise may be present during the test phase. The library and test signals can then be modeled, respectively, as

$$\mathbf{x}^{\mathcal{L}q} = \mathbf{s}^{\mathcal{L}q}, \qquad \mathbf{x} = \mathbf{s} + \mathbf{w}, \tag{5.42}$$

where $\mathbf{s}^{\mathcal{L}q} = [s^{\mathcal{L}q}(1), \dots, s^{\mathcal{L}q}(N)]^T$, $q = 1, \dots, Q$, and $\mathbf{s} = [s(1), \dots, s(N)]^T$ are the signals due to vibration and $\mathbf{w} = [w(1), \dots, w(N)]^T$ is the background noise. With $\mathbf{r}_k^{\mathcal{L}q}$, $\mathbf{r}_k$, and $\mathbf{v}_k$ being the respective BLCs of $\mathbf{s}^{\mathcal{L}q}$, $\mathbf{s}$, and $\mathbf{w}$, the BLCs of $\mathbf{x}^{\mathcal{L}q}$ and $\mathbf{x}$ can then be expressed as

$$\mathbf{z}_k^{\mathcal{L}q} = \mathbf{r}_k^{\mathcal{L}q}, \qquad \mathbf{z}_k = \mathbf{r}_k + \mathbf{v}_k. \tag{5.43}$$

In reality, both indoor and outdoor HCI applications suffer from band-limited noise. Such noise may arise from artificial sources such as road vehicles and air movement machinery including fans and ventilation or air-conditioning units [30]. The case where $\mathbf{w}$ is band-limited is therefore considered here so that there exists a set of indices I_0 where the BLCs $\mathbf{r}_k$ for $k \in I_0$ have sufficiently high local SNRs, i.e., $\mathbf{v}_k \approx \mathbf{0}$. Here $\mathbf{0}$ is a length-N vector of zeros. The correlation in Equation (5.29) can then be expressed as

$$\Psi_{\mathbf{z}_k, \mathbf{z}_k^{\mathcal{L}q}}(\ell) = \begin{cases} \Psi_{\mathbf{r}_k, \mathbf{r}_k^{\mathcal{L}q}}(\ell), & \forall k \in I_0; \\ \Psi_{\mathbf{r}_k, \mathbf{r}_k^{\mathcal{L}q}}(\ell) + \Psi_{\mathbf{v}_k, \mathbf{r}_k^{\mathcal{L}q}}(\ell), & \forall k \notin I_0. \end{cases} \tag{5.44}$$

When the noise bandwidth is narrow, a few BLCs are contaminated with noise. Since each of the correlation functions is normalized before they are summed in Equation (5.40), when the BLCs with high SNR outnumber the noisy BLCs, the contribution of the noise-free terms $\Psi_{\mathbf{r}_k, \mathbf{r}_k^{\mathcal{L}_q}}(\ell)$ dominates that of the noisy terms $\Psi_{\mathbf{v}_k, \mathbf{r}_k^{\mathcal{L}_q}}(\ell)$. As a result, $\Upsilon_q(\ell)$ is robust to noise even if the SNR of the sensor signal is low, provided the noise is sufficiently band-limited. This is not true of conventional CC-LTM described in section "Time-reversal Theory-based LTM."

5.5.3 Experiment Results

In this subsection, experiment results are presented to evaluate the perfor-mance of different LTM algorithms in the context of touch interfaces. In the following experiments, data are collected on an aluminum and a glass plate of dimensions $0.6\,\text{m} \times 0.6\,\text{m} \times 5.0\,\text{mm}$ and $0.6\,\text{m} \times 0.6\,\text{m} \times 4.0\,\text{mm}$, respectively. On the surface of each plate, a single Murata PKS1-4A1 shock sensor is mounted to capture vibration signals. Although the frequency response of the Murata sensor rolls off beyond 8 kHz, the analog sensor outputs are sampled at 96 kHz using a PreSonus FireStudio for high resolution in time. The calibration points are arranged in a 5×5 grid, as depicted in Figure 5.8. Choosing the sensor location to be the origin, i.e., $(u, v) = (0, 0)$, the lower left corner of the grid is located at $(u^{\mathcal{L}_1}, v^{\mathcal{L}_1}) = (10, 10)$ cm. The distances between adjacent columns of the grid from left to right (and adjacent rows from bottom to top) are respectively 1, 2, 3, and 4 cm.

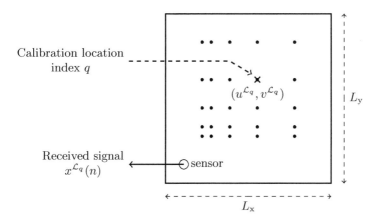

Figure 5.8 The schematic of a typical LTM setup for touch-interface applications.

During calibration, five impacts are generated at each of the pre-defined locations, resulting in a total of 125 signals. For testing, each of the collected signal takes turn being the test signal while the remaining 124 being the library. When a test signal is matched with one of the other four that correspond to the same impact location, we have an accurate match. For all experiments, the performance of each algorithm is measured by the number of accurate matches N_c out of 125 test cases,

$$C = \frac{N_c}{N_t}. \tag{5.45}$$

In what follows, the effect of the filter choice on the performance of the BLC-LTM method is first examined. Due to the frequency response of the shock sensor, it is reasonable to assume that the received signals are band-limited and denote the maximum frequency by f_B. In this experiment, we choose K filters where the passband of the k-th filter is $[f_B(k-1)/K, f_B k/K]$ so that the passbands of the filters cover the whole spectrum of the signal. For each test case, a noise limited to the band $[3, 5]$ kHz is generated and added to the clean test signal at -12 dB SNR. The resultant variation in performance of BLC-LTM for different values of K is shown in Figure 5.9(a). As previously discussed, as K increases, the BLCs that are free of noise dominate in numbers. The noise-free terms therefore outweighs the noisy terms in Equation (5.40). As a result, the localization accuracy is improved as K increases, as can be seen from Figure 5.9(a). In the following experiments,

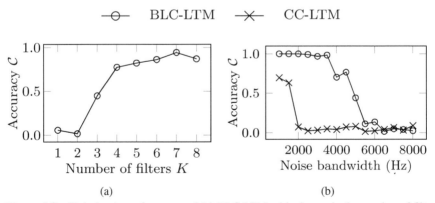

Figure 5.9 Variation in performance of (a) BLC-LTM with change in the number of filters and (b) BLC-LTM and cross-correlation-based LTM (CC-LTM) with change in the noise bandwidth with the SNR maintained at -12 dB.

we fix $K = 8$, which is equivalent to choosing filters with passbands of $[0, 1]$ kHz, $[1, 2]$ kHz, $[2, 3]$ kHz, $[3, 4]$ kHz, $[4, 5]$ kHz, $[5, 6]$ kHz, $[6, 7]$ kHz, and $[7, 8]$ kHz.

The performance of BLC-LTM is next compared with that of CC-LTM [2] at different noise bandwidths. In this experiment, test signals are generated by adding synthetic noise to each library signal while maintaining the SNR at –12 dB. The noise bandwidth is varied by generating noise within frequency bands of $[0, 1]$ kHz, $[0, 2]$ kHz, $[0, 3]$ kHz, $[0, 4]$ kHz, $[0, 5]$ kHz, $[0, 6]$ kHz, $[0, 7]$ kHz and $[0, 8]$ kHz. The experiment result, the localization accuracy of BLC-LTM and CC-LTM versus the noise bandwidth, is shown in Figure 5.9(b). As can be seen from the figure, both algorithms exhibit low accuracies for a large noise bandwidth (above 4 kHz, which is half the bandwidth of the signal received from the Murata sensor). The performance of BLC-LTM, however, increases as the noise bandwidth decreases, as opposed to the CC-LTM method. It is apparent that the conventional cross-correlation does not provide any means to suppress the effect of noise regardless of the bandwidth. On the contrary, utilizing the measure in Equation (5.40), the BLC-LTM algorithm can make use of the dominance of noise-free BLCs to achieve robustness to band-limited noise.

Finally, BLC-LTM and CC-LTM are compared with the Z-LTM [12]. As discussed in section "Classical Plate Theory-based LTM," the Z-LTM algorithm is a non-correlation-based method. Here, we evaluate the performance of the algorithms at different SNRs for two cases, white noise and band-limited noise of passband $[3, 4]$ kHz. The variation of the classification accuracies of the three algorithms on the glass plate with respect to SNR is shown in Figure 5.10(a) and 5.10(b), for the cases of white and band-limited noise, respectively. Figures 5.10(c) and 5.10(d) show the respective experiment results on the aluminum plate. It can be seen from the figures that the three algorithms perform equally well under high SNR conditions (above 0 dB) for all the cases. BLC-LTM and CC-LTM, however, exhibit higher accuracies than Z-LTM when the SNR is low. This is due to the robustness of cross-correlation-based methods to uncorrelated noise. On the contrary, non-correlation-based methods such as Z-LTM do not possess this property and have their extracted feature severely distorted when the noise energy begins to dominate the signal energy, leading to poor localization performance. While BLC-LTM can achieve the same performance as that of CC-LTM in lower SNR (by up to -15 dB) for narrow-band noise, the latter modestly outperforms the former when the noise is white. This agrees with the previous discussion on BLC-LTM's performance versus noise bandwidth.

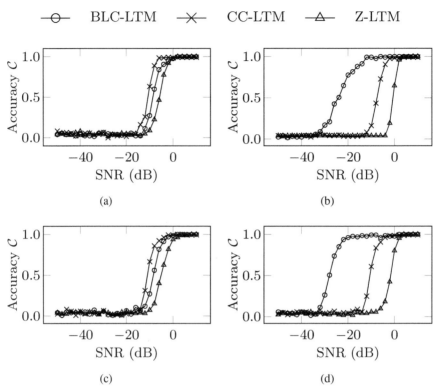

Figure 5.10   Localization accuracy vs SNR on glass for (a) white noise and (b) band-limited noise, and on aluminum for (c) white noise and (d) band-limited noise.

5.6 Concluding Remarks

In this chapter, two approaches of LTM for human–computer touch interface applications have been reviewed. In the time-reversal-theory-based approach, the signal itself is deployed as the signature for matching. On the other hand, the approach based on the classical model only extracts certain location-dependent features for utilization in matching. The latter approach requires less computation complexity since it is sufficient to store and process only partial information of each vibration signal. This, however, results in algorithms being sensitive to any signal distortion that may corrupt the feature(s) of interest. To this extent, time-reversal-based LTM is more robust due to the inclusion of the whole signal in the matching process. The BLC-LTM algorithm combines the advantages of both approaches to achieve robustness to band-limited noise.

In practice, one of the disadvantages of the LTM methods is that they can only localize impacts made at points belonging to a discrete set of pre-defined locations. To achieve high resolution in impact localization, it is therefore necessary to include as many points as possible in the calibration set. Given that features have to be extracted for every possible tap location made by the user, this incurs extensive calibration and intensive computation. In addition, the localization performance depends on the assumption of similarity between taps at the same location. This assumption, however, is often violated in reality due to inconsistencies in the style of tapping or any deviation of the test condition from that of the calibration phase.

References

[1] Paradiso, J., and Checka, N. (2002). "Passive acoustic sensing for tracking knocks atop large interactive displays," in *Proceedings of the IEEE Sensors*, Orlando, FL, 521–527.

[2] Pham, D. T., Al-Kutubi, M., Ji, Z., Yang, M., Wang, Z. B., and Catheline, S. (2005). "Tangible acoustic interface approaches," in *Proceedings of the International Virtual Conference IPROMS*, 497–502. Available at: https://www.elsevier.com/books/intelligent-production-machines-and-systems-first-i-proms-virtual-conference/pham/978-0-08-044730-8

[3] Crevoisier, A., and Polotti, P. (2005). "Tangible acoustic interfaces and their applications for the design of new musical instruments," in *Proceedings of the International Conference New Interfaces for Musical Expression*, Vancouver, BC, 97–100.

[4] Chowdhury, T., Aarabi, P., Zhou, W., Zhonglin, Y., and Zou, K. (2013). "Extended touch user interfaces," in *Proceedings 2013 IEEE International Conference Multimedia and Expo*, San Jose, CA, 1–6.

[5] Yoshida, S., Shirokura, T., Sugiura, Y., Sakamoto, D., Ono, T., Inami, M., et al. (2016). Robojockey: designing an entertainment experience with robots. *IEEE Comput. Graph. Appl.* 36, 62–69.

[6] Reju, V. G., Khong, A. W. H., and Sulaiman, A. B. (2013). Localization of taps on solid surfaces for human-computer touch interfaces. *IEEE Trans. Multimed.* 15, 1365–1376.

[7] Ing, R. K., Quieffin, N., Catheline, S., and Fink, M. (2005). In solid localization of finger impacts using acoustic time-reversal process. *Appl. Phys. Lett.* 87, 1–3.

[8] Knapp, C., and Carter, G. C. (1976). The generalized correlation method for estimation of time delay. *IEEE Trans. Acoust. Speech Signal Process.* 24, 320–327.

[9] Ziola, S. M., and Gorman, M. R. (1991). Source location in thin plates using cross-correlation. *J. Acoust. Soc. Am.* 90, 2551–2556.

[10] Arun, K. R., Ong, E., and Khong, A. W. H. (2011). "Source localization on solids using Kullback-Leibler discrimination information," in *Proceedings of the 8th International Conference Information, Communications and Signal Processing (ICICS)*, Singapore, 1–5.

[11] Poletkin, K., Yap, X., and Khong, A. W. H. (2010). "A touch interface exploiting the use of vibration theories and infinite impulse response filter modeling based localization algorithm," in *Proceedings of the IEEE International Conference Multimedia and Expo (ICME)*, Singapore, 286–291.

[12] Arun, K. R., Yap, X., and Khong, A. W. H. (2011). A touch interface exploiting time-frequency classification using Zak transform for source localization on solids. *IEEE Trans. Multimed.* 13, 487–497.

[13] Nguyen, Q. H., Reju, V. G., and Khong, A. W. H. (2015). "Noise robust source localization on solid surfaces," in *Proceedings of the 10th International Conference Information Communications and Signal Processing (ICICS)*, Singapore, 1–5.

[14] Fink, M. (1992). Time reversal of ultrasonic fields – Part I: Basic principles. *IEEE Trans. Ultrason. Ferroelectr. Freq. Control* 39, 555–566.

[15] Ing, R. K., Quieffin, N., Catheline, S., and Fink, M. (2005). Tangible interactive interface using acoustic time reversal process. *J. Acoust. Soc. Am.* 117, 2560–2560.

[16] Ribay, G., Clorennec, D., Catheline, S., Fink, M., Kirkling, R., and Quieffin, N. (2005). Tactile time reversal interactivity: Experiment and modelization. In *Proceedings of the IEEE International Ultrasonics Symposium*, Rotterdam, 2104–2107.

[17] Ribay, G., Catheline, S., Clorennec, D., Ing, R. K., Quieffin, N., and Fink, M. (2007). Acoustic impact localization in plates: Properties and stability to temperature variation. *IEEE Trans. Ultrason. Ferroelectr. Freq. Control* 54, 378–85.

[18] Yang, M., Pham, D., Al-Kutubi, M., Ji, Z., and Wang, Z. (2009). "A new computer interface based on in-solid acoustic source localization," in *Proceedings of the IEEE International Conference Industrial Informatics*, Cardiff, 131–135.

[19] Bai, M. R., and Tsai, Y. K. (2011). Impact localization combined with haptic feedback for touch panel applications based on the time-reversal approach. *J. Acoust. Soc. Am.* 129, 1297–1305.

[20] Graff, K. F. (1991). *Wave Motion in Elastic Solids*. Mineola, NY: Dover Publications.

[21] Ventsel, E., and Krauthammer, T. (2001). *Thin Plates and Shells: Theory, Analysis, and Applications*. Boca Raton, FL: CRC Press.

[22] Szilard, R. (2004). *Theories and Applications of Plate Analysis: Classical, Numerical and Engineering Methods*. Hoboken, NJ: Wiley.

[23] Mindlin, R. D. (1951). The influence of rotatory inertia and shear on the flexural motions of isotropic elastic plates. *J. Appl. Mech. Trans. ASME.* 18, 31–38.

[24] Viktorov, I. A. (1967). *Rayleigh and Lamb Waves: Physical Theory and Applications*. New York, NY: Plenum Press.

[25] Su, Z., and Ye, L. (2009). "Fundamentals and Analysis of Lamb Waves," in *Identification of Damage Using Lamb Waves*. London: Springer, 48, 15–58.

[26] Cohen, L. (1989). Time-frequency distributions – A review. *Proc. IEEE* 77, 941–981.

[27] Janssen, A. (1988). The Zak transform: a signal transform for sampled time-continuous signals. *Philips J. Res.* 43, 23–69.

[28] Bolcskei, H., and Hlawatsch, F. (1997). Discrete Zak transforms, polyphase transforms, and applications. *IEEE Trans. Signal Process.* 45, 851–866.

[29] O'Hair, J. R., and Suter, B. W. (1996). The Zak transform and decimated time-frequency distributions. *IEEE Trans. Signal Process.* 44, 1099–1110.

[30] Berglund, B., Hassmen, P., and Job, R. F. S. (1996). Sources and effects of low-frequency noise. *J. Acoust. Soc. Am.* 99, 2985–3002.

6

Automatic Placental Maturity Grading via Deep Convolutional Networks

Baiying Lei[1], Feng Jiang[1], Yuan Yao[2], Wanjun Li[1], Siping Chen[1], Dong Ni[1] and Tianfu Wang[1]

[1] National-Regional Key Technology Engineering Laboratory for Medical Ultrasound, Guangdong Key Laboratory for Biomedical Measurements and Ultrasound Imaging, School of Biomedical Engineering, Health Science Center, Shenzhen University, Shenzhen, China
[2] Department of Ultrasound, Affiliated Shenzhen Maternal and Child Healthcare, Hospital of Nanfang Medical University, Shenzhen, China

Abstract

The accuracy of placental maturity staging is important to the clinical diagnosis of small gestational age, stillbirth, and fetal death. Fetal viabilities, various gestational ages, and complicated imaging process have made placental maturity evaluation a tedious and time-consuming task. Despite development of numerous tools and techniques to access placental maturity, automatic placental maturity remains challenging. To address this issue, we proposed to automatically grade placental maturity by obtaining gray-scale features from B-mode ultrasound and vascular blood flow information from color Doppler energy images based on deep convolutional neural networks, such as AlexNet and VGGNet. The deep network can achieve high-level features to further improve the accuracy of placental maturity grading. Both the transfer learning strategy and unique data augmentation technique are utilized to further boost the recognition performance. Extensive experimental results demonstrate that our method achieves promising performance in placental maturity evaluation and would be beneficial in clinical application.

6.1 Introduction

Ultrasound (US) imaging has been widely applied in prenatal diagnosis and prognosis due to its benefits such as non-radiation, direct-use, and low cost [1–8]. B-mode gray-scale US (BUS) images have been widely used to detect placental abnormalities (e.g., miscarriage, fetal death, still birth, small gestational age, and various pregnancy complications). It's important to detect the structure and function of placenta using prenatal US [9, 10]. The placenta is an important organ for fetal placental blood circulation, gas exchange, nutrient supply, and fetal waste elimination. Therefore, the placental function is critical to the development of fetus and normal pregnancy. Accurate and timely placental function evaluation will provide effective and reliable information for clinical diagnosis of fetal and maternal abnormalities and diseases [11, 12]. Placental maturity is one of the most important parameters of the placenta for the objectivity and accuracy of the grading results. In general, accurate assessment of the placental maturity stage requires knowledge and experience of the physician and the radiologist.

Existing algorithms for placental maturity classification are mainly based on B-mode US images. In 1979, Grannum [13] proposed to divide the placenta maturity into four levels based on the gray images. Doctors collected B-mode US images, observed and then analyzed the placental calcification and blood in order to make a final decision. However, placental maturity evaluation remains a challenging issue, and automatic computer-assisted diagnosis would be desirable as it not only reduces errors caused by judgment [14], but also provides an attractive and meaningful standardization tool [15]. Errors pertaining to classification results will lead to misdiagnosis of placenta-related diseases. Therefore, it is necessary to find a more objective and automatic classification method [16].

Some automatic methods have been invented for placental assessment. In [17], Lei et al. proposed a placental evaluation based on Fisher vector (FV) and invariant descriptor [18], and achieved promising grading performance. In [19], Li et al. proposed to use a dense descriptor such as DAISY to address the placental evaluation problem. This dense sampling method outperformed the traditional co-variant affine detectors such as Harris, Hessian, and multi-scale Harris. Although good performance in placental evaluation has been achieved with B-mode gray-scale US images, it was argued that this method is limited due to the lack of blood flow information. Blood vessels play an important role in placental function evaluation as non-branched and branched blood vessels are essential in morphogenesis during pregnancy [20–22].

The placental angiogenesis inhibition would lead to fetal growth restriction in the development of the placenta. Meanwhile, it is known that color Doppler energy (CDE) imaging is a technique to obtain the velocity distribution of blood flow in the human body tissue plane in the form of gray scale or color bands [23, 24]. This technique can be adopted to understand not only the structure of human tissue, but also the kinematic information of the body's blood flow (or organization). In Doppler imaging, the vessels that are traveling within the region of interest can form the flow spectrum shape and provide flow information [25]. Since there are a large number of visible blood vessels within the placenta after 14 weeks of pregnancy, placental blood vessel detection and blood flow information are particularly important for fetal placental development and evaluation [26]. Therefore, exploring blood flow using CDE can enhance placental evaluation. However, this information has often been ignored or unconsidered in previous studies. Table 6.1 illustrates the benchmark of placental maturity based on B-mode image and CED image [27, 28].

6.2 Related Work

Various methods are proposed in the literature to address above-mentioned problems. One of the most common ways is to use low-level hand-crafted features (i.e., SIFT, Haar and HoG) as an image descriptor to represent the images. The low-level feature is further encoded by conventional methods such as the bag of visual words (BoVW) [29], vector of locally aggregated descriptor (VLAD) [30, 31], and FV [32, 33] to improve the effectiveness of recognition. However, existing handcrafted features extracted from consecutive 2D US images are still unsatisfactory for placental maturity grading. These methods employ machine learning which is inferior to the prevalent deep learning methods. In 1998, Yann LeCun [34] first proposed the deep learning method to deal with problems of image recognition. An image of size 32×32 is used as input and the algorithm outputs a number corresponding to the picture via a seven-layer neural network excluding the input layer (i.e., LeNet-5). The first five layers of LeNet-5 consist of convolutional layers and down-sampling layers. The last two layers are the fully connected (FC) layer and the output layer. Convolutional layers adopt a 5×5 kernel to enhance the original signal and reduce noise. Down-sampling layers reduce the amount of processed data and retain the useful information by sampling. LeNet-5 is not good at handling images with rich pixels or large size.

Table 6.1 Placental maturity characteristics and representative images

Characteristic	Stage 0	Stage I	Stage II	Stage III
Chorionic plate	Straight, smooth and Chiseled	Slight undulating	In a serrated form, may extend into the substance of the placenta, but not the basal layer	Jaggered, stretched into basal layer
Substance	Uniform	Unvenly distributed, scattered, point-like	Linearly echogenic, comma-like densities	Circular densities, halo with cast acoustic shadow
Basal Layer	No echo	No echo	Linear aligned, point-like echo	Large, confluent with basal layer, acoustic shadow
BUS image				
CDE image				

In 2006, Hinton [35] published a paper titled "Reducing the dimensionality of data with neural networks," which led to deep learning becoming a hot topic in various applications and research again. In 2012, Krizhevsky et al. utilized the classic convolutional neural network (CNN), namely, AlexNet, to achieve remarkable results in natural image classification [36]. Compared to LeNet-5, AlexNet doesn't change a lot in the structure and has the following advantages:

1. Data resource: Li et al. [37] proposed to use the dataset of ImageNet with 1000 classes of labeled natural images to provid data support for AlexNet.
2. Computational resource: As a high-speed parallel and powerful computing accessory, GPU provides hardware support for AlexNet in calculation and improvement.
3. Structural development: Compared to LeNet-5, AlexNet increases layers of the network, adopts ReLU activation function and sets the discard layer. These optimized algorithms are critical to the success of AlexNet.

The promising performance achieved by CNNs promotes successful application of various enhanced data structures, such as ZFNet [38], VGGNet [39], GoogleNet [40], and ResNet [41]. ZFNet made a breakthrough to provide characteristic visualization method. It uses a smaller kernel in the first layer to retain more original image information. VGGNet only employs a 3×3 kernel in the entire convolutional process. The stride of the down-sampling layers is 2 pixels. The network has a deeper structure which goes up to 19 layers. GoogleNet presents a 22-layer network structure and ResNet introduces residual learning. These improvements greatly boost the accuracy of image classification. However, the complex network may lead to over-fitting, network degradation, adjustment difficulties, and increased data requirement due to many parameters. Convolutional neural network with high accuracy needs to be explored in the medical image application [42]. Motivated by this, we explore deep neural network to stage placental maturity with transfer learning strategy. Specifically, we exploit different depth neural network models pretrained from large-scale well-annotated datasets. We combine the techniques of data augmentation and transfer learning in the deep neural network models. To the best of our knowledge, this is the first automatic grading method using CNN, which has great potential in the practical application of routine US examination and prenatal care.

6.3 Methodology

6.3.1 Convolutional Neural Network

Convolutional neural network introduces convolutional calculation and down-sampling in the hierarchical structure based on traditional neural networks. Local features of images are better extracted via convolutional calculation, and the down-sampling process can generate convolved features with spatial invariance.

Figure 6.1 shows the basic structure of AlexNet. The structure of AlexNet has eight layers, including five convolutional layers and three FC layers. Each convolutional layer is linked to a down-sampling layer. The convolutional layers are used to extract features and reduce noise. The down-sampling layer reduces computational cost and maintains spatial consistency between the front and rear images. The FC layers are utilized to classify images. This section describes each part of the CNN in detail according to the structure of AlexNet, mainly including convolution, down-sampling, and activation function.

6.3.1.1 Convolution

Figure 6.2 summarizes the one-dimensional convolutional dot product, that is, $y = x \times w$. Traditional neural networks require all neurons to be involved in the weighted computation to obtain an output. But CNN only requires local neurons to compute the weights and this is known as local receptive field. Using local receptive field can reduce many parameters. If sharing weights are employed in subsequent calculation, less parameters will be needed. Local features are expressed effectively via local receptive fields and sharing parameters. The high-level features can be achieved by increasing the depth of the network.

The two-dimensional convolution is similar to the process where a filter is used to deal with the image, that is, the sum of the corresponding elements is

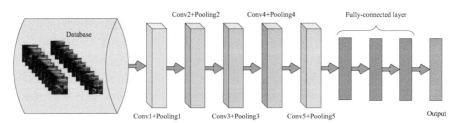

Figure 6.1   Structure of AlexNet.

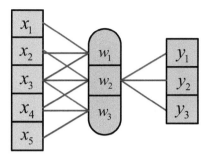

Input   Convolutional kernel   Output

Figure 6.2   Diagram of one-dimensional convolution.

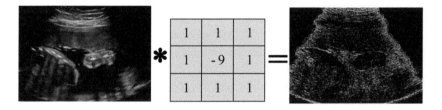

Figure 6.3   Diagram of two-dimensional convolution.

multiplied. Different convolutional kernels may produce completely different results. The computing results are shown in Figure 6.3. Image convolution is usually three-dimensional convolution, and the inner product between the input image and the convolutional kernel is calculated. Features of the image are extracted by the movement of the convolutional kernel in the image plane. The convolutional kernel size will also affect the extraction of the local features to some extent. If the information of the image boundary is important, it is necessary to complement zero in the surrounding area to preserve the characteristics of the boundary. In the convolutional layer, we need to identify the number and the size of the kernel, the weight in the kernel, and the stride of computing convolution. We also should know whether to complement zero at the image boundary.

6.3.1.2 Down-sampling

Due to the large number of neurons and layers, the number of parameters in CNN is too high, which increases the need for computing resources and long operation time. In order to avoid the huge resource consumption and

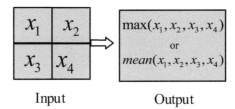

Input   Output

Figure 6.4   Diagram of 2 × 2 down-sampling.

time-consuming training, down-sampling is followed by extracting features using convolutional layers to reduce the computational cost while ensuring the spatial consistency of the image calculation. Down-sampling, also known as pooling, takes the average or maximum of the entire area as the output. Figure 6.4 illustrates the process.

Down-sampling takes the neighborhood information of the target point into the statistical process to keep the image relatively constant so that the output is similar to the original data. Moreover, down-sampling can vastly reduce the dimension of the feature data and the possibility of over-fitting in training.

6.3.1.3 Activation functions

Activation functions are usually designed to solve non-linear questions. Thus, the activation functions are usually non-linear. Since the common optimization is gradient-based, it must be differentiable. The monotonous activation function ensures that the single-layer network is convex. The range of output values is also infinite. Some of the common activation functions are sigmoid, tanh, and ReLU.

(1) Sigmoid: Sigmoid is a non-linear function denoted as below:

$$f(x) = \frac{1}{1 + e^{-x}} \tag{6.1}$$

The range of the output is from 0 to 1. Sigmoid has some shortcoming: when the input is close to both ends (i.e., too large or too small), the gradient reaches zero. Therefore, if the initial value is set incorrectly, it will make the network difficult to learn. The average output isn't zero. The calculated weight of the gradient is always positive when the input is positive and the data is too hard to adjust.

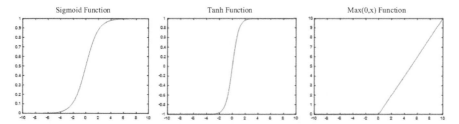

Figure 6.5 Curve of three functions. From left to right, sigmoid, tanh and ReLU.

(2) Tanh: Tanh is similar to sigmoid function, and is formulated as:

$$\tanh(x) = 2 \, sig \, \text{mod}(2x) - 1. \tag{6.2}$$

Since tanh has the mean of zero, it is much easier to train compared to sigmoid.

(3) ReLU: The output of ReLU is non-negative, formulated as:

$$f(x) = \max(0, x). \tag{6.3}$$

The study shows that ReLU converges fast and does not require complicated calculation. It is easy to never be activated again for the high learning efficiency. To address it, there exist many improved functions. Figure 6.5 illustrates these three functions.

6.3.2 Automatic Grading Algorithm Based on CNN

6.3.2.1 Framework

In this study, we explore three kinds of CNN according to their layers which have 8 layers, 16 layers, and 19 layers, respectively. First, we adopt the data augmentation method to expand the training dataset. Then, transfer learning strategy is used to avoid over-fitting during the training process. Specifically, the parameters extracted from well-pre-trained networks are migrated to our network. Finally, we utilize the expanded dataset to train the network to achieve the grading accuracy. Figure 6.6 shows the framework of our proposed method.

6.3.2.2 Data augmentation

Since the training of CNN requires abundant data, and the amount of the placental image data is too small, it is difficult to adjust the network. Common

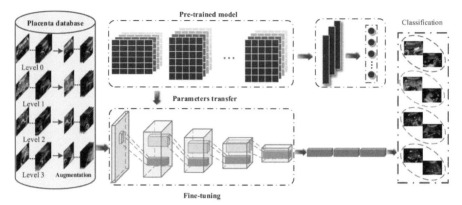

Figure 6.6 Framework of our proposed method.

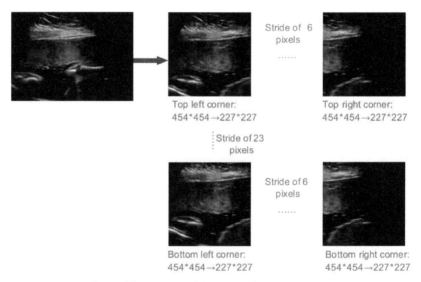

Figure 6.7 Process of data expansion and cropping.

methods of data augmentation include image flip, rotation, crop, zoom, twist, etc. In order to ensure the relative integrity of placenta in image and the effectiveness of training, we crop the images using the certain stride and zoom the images to achieve data expansion. The process of data augmentation is shown in Figure 6.7. From the figure, we obtain an image of size 454×454 with a transversal stride of 6 pixels and a longitudinal stride of 23 pixels. After 0.5 times zoom transformation, an obtained image of size 227×227 is used as the input image of the network.

6.3.2.3 Transfer learning

Transfer learning refers to the network trained by the large-scale data for initializing small amount of data. This approach is widely applied in problems of large correlation. Machines similar to humans can think and learn quickly and acquire new knowledge. Transfer learning is divided into sample transfer, feature transfer, model transfer, and relational transfer according to the transferred parts. Among them, sample transfer means that similar samples are utilized to weight rich learning content. Feature transfer extracts features via the same feature extraction methods during the extraction of features. Model transfer is known as the existing structure of network and samples to tune parameters. Relational transfer intends to integrate the relationship among the existing learning concepts into own problems. Due to the certain similarity of image classification in the nature and finite data, we use convergent parameters of trained networks as initial values of our network. It can reduce the possibility of under-fitting caused by small datasets.

6.3.2.4 Feature visualization

Convolutional neural network is known as the black boxes. We can only control the input, and then achieve the results with the unknown internal learning information. Feature visualization provides us with a deeper understanding of the training process of CNN. In the forward propagation and backward propagation, feature visualization can intuitively observe changes in characteristics, and we can learn better tuning methods. The following figures are the results of feature visualization based on placental images of level 0. Figure 6.8 shows the visualization results of the five convolutional layers of AlexNet, VGG-F, VGG-S, VGG-M, respectively.

6.4 Experiments

6.4.1 Experimental Settings

Our data are collected from Shenzhen Maternal and Child Health Hospital for deep learning experiments. The obtained data are used to assess the effects of the existing networks in placental maturity grading, primarily for training and validation. Specifically, 90% of the data are employed for training, the rest are used for validation. Subsequently, we collected the placental data of multiple machines from Shenzhen Maternal and Child Health Hospital, including GE VOLUSON E8 (16 images of grade 0, 10 images of grade I, 14 images of grade II, 5 images of grade III), SAMSUNG ACCUVIX A (24 images of

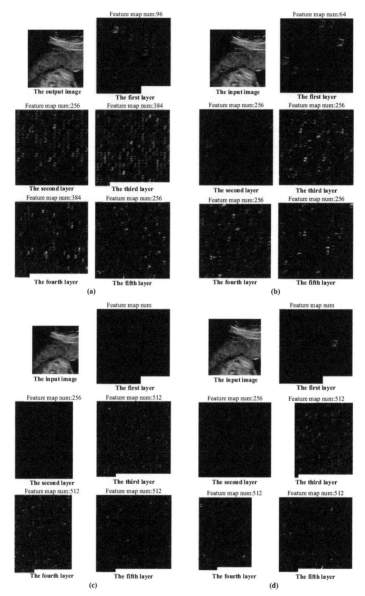

Figure 6.8   Feature visualization results of five convolutional layers. (a–d) Feature maps of AlexNet, VGG-F, VGG-S and VGG-M, respectively.

grade 0, 19 images of grade I, 10 images of grade II, 4 images of grade III), SIEMENS ACUSON S2000 (30 images of grade 0, 12 images of grade I, 12 images of grade II, 8 images of grade III), SAMSUNG WS80A (18 images of grade 0, 13 images of grade I, 15 images of grade II, 7 images of grade III). Collected images are still anterior wall of placenta as it is better to observe the whole placenta. The networks in the experiments are the existing networks mentioned above.

6.4.2 Experimental Results

AlexNet, VGG-F, VGG-S, VGG-M, VGG-16, and VGG-19 were for training and validation, respectively. The results of AlexNet listed in Figure 6.9(a) illustrate that the results of validation reach convergence after training 12

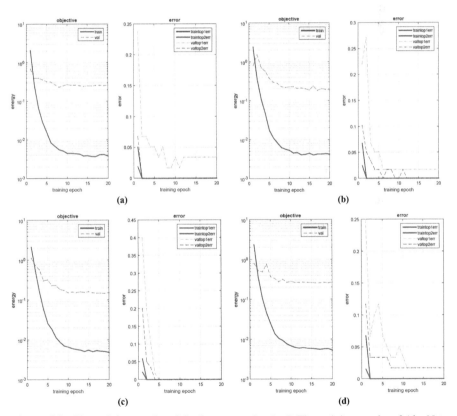

Figure 6.9 The training results of the four networks. (a–d) The training results of AlexNet, VGG-F, VGG-S, and VGG-M, respectively.

times. The error rate is 0.034. Compared with the previous method, the achieved results increase by 3.4% in the precision rate. Figure 6.9(b–d) summarizes that the error rate of VGG-F, VGG-S, and VGG-M is 0.017, 0, 0.017, respectively. Moreover, the rapid convergence of these networks also proves that transfer learning saves computational time and resource effectiveness in automatic placental maturity grading. We will use the new test dataset to further verify the feasibility of automatic placental maturity grading via deep learning.

The network with many layers may be able to get better results, but the speed is quite low. The network with less layers has been able to get great results for less complex classification problems. Figure 6.9 shows that the error rate of VGG-16 is 0.0085 and reveals that the error rate of VGG-19 is 0.017.

We use up-to-date collected data to test trained networks including AlexNet, VGG-F, VGG-S, VGG-M, VGG-16, and VGG-19. Table 6.2 shows the test results of GE VOLUSON E8, SAMSUNG ACCUVIX A, SIEMENS ACUSON S2000, and SAMSUNG WS80A, respectively.

6.4.3 Result Analysis

In this study, we have used four US instruments including GE VOLUSON E8, SAMSUNG ACCUVIX A, SIEMENS ACUSON, and SAMSUNG WS80A to test the trained networks. Different results are obtained due to the large differences of image quality among different types of US instruments and various shapes of placenta. In general, the CNN algorithm is promising for automatic placental maturity grading. As shown in Table 6.2, we can see that CNN obtains the best grading performance among different test datasets, which demonstrates the effectiveness and robustness of CNN in placental maturity grading. It is found that deep networks can achieve better staging effects than the shallow networks using different networks to grade placenta maturity automatically. However, the deep networks normally are more time-consuming than the shallow networks, which is also illustrated in Table 6.2. In each test dataset, VGG-16 and VGG-19 obtain the best performance for their deep network structures. Furthermore, Figure 6.9 shows the training results of the four networks including AlexNet, VGG-F, VGG-M, and VGG-S. In general, the structure of VGG-S is more complex than others. It is seen that the error of the validation set in VGG-S is much lower than that in AlexNet when both reach convergence, which means that VGG-S is trained more precisely. These findings are consistent with those in Figure 6.8. Compared

Table 6.2 Test results of different ultrasonic machines using different networks

Network	GE VOLUSON E8				SAMSUNG ACCUVIX A				SIEMENSACUSON S2000				SAMSUNG WS80A			
	MAP	SEN	SPE	ACC	MAP	SEN	SPE	ACC	MAP	SEN	SPE	ACC	MAP	SEN	SPE	ACC
AlexNet	0.92	0.93	0.98	0.95	0.94	0.94	0.96	0.97	0.95	0.91	0.94	0.93	0.93	0.93	0.99	0.97
VGG-F	0.92	0.93	0.98	0.93	0.97	0.92	0.99	0.97	0.95	0.92	0.94	0.93	0.93	0.95	0.99	0.97
VGG-S	0.92	0.94	0.95	0.98	0.95	0.95	0.98	0.94	0.97	0.92	0.96	0.97	0.93	0.91	0.98	0.97
VGG-M	0.93	0.94	0.98	0.96	0.96	0.92	0.97	0.92	0.93	0.92	0.98	0.94	0.93	0.92	1	0.93
VGG-16	0.93	0.93	0.99	0.96	0.94	0.97	0.92	0.97	0.91	0.91	0.94	0.93	0.93	0.92	1	0.97
VGG-19	0.92	0.92	0.98	0.97	0.94	0.96	0.97	0.93	0.93	0.92	0.98	0.95	0.93	0.92	0.98	0.93

with other three networks, the feature maps of VGG-S are more specific and discriminative for placental maturity grading. These high-level features are only extracted from deep networks, while low-level features are obtained from shallow networks, which affects the classification performance. In our future work, we will design a faster and more accurate network to improve the efficiency and accuracy of automatic placental maturity grading.

6.5 Conclusion

In this study, we proposed a new method for placental maturity staging based on both BUS and CDE images. We utilized different depth CNN models with transfer learning strategy for automatic placental maturity grading. We tested the data of multiple machines into the deep learning framework to prove the generalization of our method. The experimental results demonstrate the powerful classification ability of CNN for solving placental detection problem. Both CNN-8 and CNN-16 have achieved impressive performance in staging. The high placental maturity grading performance via CNN models and transfer learning shows great potential for clinical applications. For our future work, we will collect larger dataset for CNN model testing, and exploit the relations between the noise of training dataset and CNN to improve placental maturity staging performance. Namely, (1) data expansion: the size of the data has a significant effect on the network. In the next experiment, we aim to improve our experiments further via rich machine types and data sizes; (2) more reasonable experiments: since parameters of the network were not tuned in this experiment. In our future work, the data collected by the new machines can be added to the validation dataset, which can improve the difference tolerance of the network on similar data and improve the grading results; (3) design specific network. Although the existing networks have achieved good results, the training and testing time of deep network are too long while the shallow networks can't obtain great results. The direction of our next experiment is to set more suitable networks for this problem.

References

[1] Chang, H., Chen, Z., Huang, Q., Shi, J., and Li, X. (2015). Graph-based learning for segmentation of 3D ultrasound images. *Neurocomputing* 151, 632–644.

[2] Chen, H., Ni, D., Qin, J., Li, S., Yang, X., Wang, T., et al. (2015). Standard plane localization in fetal ultrasound via domain transferred deep neural networks. *IEEE J. Biomed. Health Informat.* 19, 1627–1636.

[3] Cheng, J.-Z., Ni, D., Chou, Y.-H., Qin, J., Tiu, C.-M., Chang, Y.-C., et al. (2016). Computer-aided diagnosis with deep learning architecture: applications to breast lesions in US images and pulmonary nodules in CT scans. *Sci. Rep.* 6:24454.

[4] D'Hooge, J., Heimdal, A., Jamal, F., Kukulski, T., Bijnens, B., Rademakers, F., et al. (2000). Regional strain and strain rate measurements by cardiac ultrasound: principles, implementation and limitations. *Eur. J. Echocardiogr.* 1, 154-170.

[5] Huang, Q., Huang, Y., Hu, W., and Li, X. (2015). Bezier interpolation for 3-D freehand ultrasound. *IEEE Trans. Hum. Machine Syst.* 45, 385–392.

[6] Huang, Q., Xie, B., Ye, P., and Chen, Z. (2015). Correspondence – 3-D ultrasonic strain imaging based on a linear scanning system. *IEEE Trans. Ultras. Ferroelectr. Frequency Control* 62, 392–400.

[7] Kellow, Z. S., and Feldstein, V. A. (2011). Ultrasound of the placenta and umbilical cord: a review. *Ultrasound Q.* 27, 187–197.

[8] Moran, M., Mulcahy, C., Daly, L., Zombori, G., Downey, P., and McAuliffe, F. M. (2014). Novel placental ultrasound assessment: Potential role in pre-gestational diabetic pregnancy. *Placenta* 35, 639–644.

[9] Stanciu, S. G., Xu, S., Peng, Q., Yan, J., Stanciu, G. A., Welsch, R. E., et al. (2014). Experimenting liver fibrosis diagnostic by two photon excitation microscopy and Bag-of-Features image classification. *Sci. Rep.* 4:4636.

[10] Tamaki, T., Yoshimuta, J., Kawakami, M., Raytchev, B., Kaneda, K., Yoshida, S., et al. (2013). Computer-aided colorectal tumor classification in NBI endoscopy using local features. *Med. Image Anal.* 17, p. 78, 2013.

[11] Li, W., Yao, Y., Ni, D., Chen, S., Lei, B., and Wang, T. (2016). Placental maturity evaluation via feature fusion and discriminative learning," in *Proceedings 2016 IEEE 13th International Symposium on Biomedical Imaging (ISBI)*, Prague, 783–786.

[12] Lei, B., Tan, E.-L., Chen, S., Li, W., Ni, D., Yao, Y., et al. (2017). Automatic placental maturity grading via hybrid learning. *Neurocomputing* 223, 86–102.

[13] Grannum, P. A. T., Berkowitz, R. L., and Hobbins, J. C. (1979). The ultrasonic changes in the maturing placenta and their relation to fetal pulmonic maturity. *Am. J. Obstetr. Gynecol.* 133, 915–922.

[14] Dubiel, M., Breborowicz, G. H., Ropacka, M., Pietryga, M., Maulik, D., and Gudmundsson, S. (2005). Computer analysis of three-dimensional power angiography images of foetal cerebral, lung and placental circulation in normal and high-risk pregnancy. *Ultrasound Med. Biol.* 31, 321–327.

[15] Goldenberg, R. L., Gravett, M. G., Iams, J., Papageorghiou, A. T., Waller, S. A., Kramer, M., et al. (2012). The preterm birth syndrome: issues to consider in creating a classification system. *Am. J. Obstetr. Gynecol.* 206, 113–118.

[16] Chou, M. M., Ho, E. S., and Lee, Y. H. (2000). Prenatal diagnosis of placenta previa accreta by transabdominal color Doppler ultrasound. *Ultrasound obstetr. Gynecol.* 15, 28–35.

[17] Lei, B., Li, X., Yao, Y., Li, S., Chen, S., Zhou, Y., et al. (2014). "Automatic grading of placental maturity based on LIOP and fisher vector," in *Proceedings of 2014 36th Annual International Conference of the IEEE Engineering in Medicine and Biology Society*, Chicago, IL, 4671–4674.

[18] Luenberger, D. G. (1978). Time-invariant descriptor systems. *Automatica* 14, 473–480.

[19] Li, X., Yao, Y., Ni, D., Chen, S., Li, S., Lei, B. et al. (2014). Automatic staging of placental maturity based on dense descriptor. *Biomed. Mater. Eng.* 24, 2821.

[20] Bude, R. O., and Rubin, J. M. (1996). Power Doppler sonography. *Radiology* 200, 21–23.

[21] Guiot, C., Gaglioti, P., Oberto, M., Piccoli, E., Rosato, R., and Todros, T. (2008). Is three-dimensional power Doppler ultrasound useful in the assessment of placental perfusion in normal and growth-restricted pregnancies? *Ultrasound Obstet Gynecol.* 31, 171–176.

[22] Burton, G. J., Charnockjones, D. S., and Jauniaux, E. (2009). Regulation of vascular growth and function in the human placenta. *Reproduction* 138, 895–902.

[23] Ozcan, T., and Pressman, E. K. (2008). Imaging of the placenta. *Ultrasound Clin.* 3, 13–22.

[24] Elsayes, K. M., Trout, A. T., Friedkin, A. M., Liu, P. S., Bude, R. O., Platt, J. F. et al. (2009). Imaging of the placenta: a multimodality pictorial review. *Radiographics* 29, 1371–1391.

[25] Lei, B., Li, W., Yao, Y., Jiang, X., Tan, E.-L., Qin, J., et al. (2017). Multimodal and multi-layout discriminative learning for placental maturity staging. *Pattern Recogn.* 63, 719–730.

[26] Guerriero, S., Ajossa, S., Lai, M. P., Risalvato, A., Paoletti, A. M., and Melis, G. B. (1999). Clinical applications of colour Doppler energy imaging in the female reproductive tract and pregnancy. *Hum. Reprod. Update* 5, 515–529.

[27] Ribeiro, R. T., Marinho, R. T., and Sanches, J. M. (2013). Classification and staging of chronic liver disease from multimodal data. *IEEE Trans. Biomed. Eng.* 60, 1336–1344.

[28] Linares, P. A., Mccullagh, P. J., Black, N. D.,and Dornan, J. (2004). "Feature selection for the characterization of ultrasonic images of the placenta using texture classification," in *IEEE International Symposium on Biomedical Imaging: Nano To Macro*, Vol. 2, Arlington, VA, 1147–1150.

[29] Lazebnik, S., Schmid, C., and Ponce, J. (2006). Beyond bags of features: spatial pyramid matching for recognizing natural scene categories," in *Proceedings of 2006 IEEE Computer Society Conference on Computer Vision and Pattern Recognition*, New York, NY, 2169–2178.

[30] Jegou, H., Perronnin, F., Douze, M., Sánchez, J., Perez, P., Schmidet, C., et al. (2012). Aggregating local image descriptors into compact codes. *IEEE Trans. Pattern Anal. Mach. Intelli.* 34, 1704–1716.

[31] Lei, B., Chen, S., Dong, N., and Wang, T. (2016). Discriminative learning for Alzheimer's disease diagnosis via canonical correlation analysis and multimodal fusion. *Front. Aging Neurosci.* 8:77.

[32] Sánchez, J., Perronnin, F., Mensink, T., and Verbeek, J. (2013). Image classification with the fisher vector: theory and practice. *Int. J. Comput. Vis.* 105, 222–245.

[33] Lei, B., Yao, Y., Chen, S., Li, S., Li, W., Ni, D., et al. (2015). Discriminative learning for automatic staging of placental maturity via multi-layer fisher vector. *Sci. Rep.* 5:12818.

[34] Lecun, Y., Bottou, L., Bengio, Y., and Haffner, P. (1998). Gradient-based learning applied to document recognition. *Proc. IEEE* 86, 2278–2324.

[35] Hinton, G. E., and Salakhutdinov, R. R. (2006). Reducing the dimensionality of data with neural networks. *Science* 313, 504–507.

[36] Krizhevsky, A., Sutskever, I., and Hinton, G. E. (2012). "ImageNet classification with deep convolutional neural networks," in *Proceedings of the 25th International Conference on Neural Information Processing Systems*, (Red Hook, NY: Curran Associates, Inc), 1097–1105.

[37] Deng, J., Dong, W., Socher, R., L. Li, J., Li, K., and Li, F. F. (2009). ImageNet: a large-scale hierarchical image database," in *Proceedings of*

2009 IEEE Conference on Computer Vision and Pattern Recognition, Miami, FL, 248–255.

[38] Zeiler, M. D,, and Fergus, R. (2013). "Visualizing and understanding convolutional networks," in *Computer Vision – ECCV 2014. Lecture Notes in Computer Science*, Vol. 8689, eds D. Fleet, T. Pajdla, B. Schiele, and T. Tuytelaars (Cham: Springer), 818–833.

[39] Simonyan, K., and Zisserman, A. (2014). Very deep convolutional networks for large-scale image recognition. arXiv preprint arXiv:1409.1556.

[40] Szegedy, C., Liu, W., Jia, Y., Sermanet, P., Reed, S., Anguelov, D., et al. (2015). "Going deeper with convolutions," in *Proceedings of Computer Vision and Pattern Recognition*, Piscataway, NJ.

[41] Lei, B., Tan, E. L., Chen, S., Zhuo, L., Li, S., Ni, D. et al. (2014). Automatic recognition of fetal facial standard plane in ultrasound image via fisher vector. *PLOS ONE* 10:e0121838.

[42] Wu, L., Cheng, J. Z., Li, S., Lei, B., Wang, T., and Ni, D. (2017). FUIQA: fetal ultrasound image quality assessment with deep convolutional networks. *IEEE Trans. Cybernet.* 47, 1336–1349.

Index

About the Editors

Andy Khong is currently an Associate Professor in the School of Electrical and Electronic Engineering, Nanyang Technological University, Singapore. He obtained his Ph.D. from Imperial College London and his B.Eng. from Nanyang Technological University, Singapore. His postdoctoral research involved the development of signal processing algorithms for acoustic microphone array and seismic sensors in perimeter security systems. His Ph.D. research was mainly on adaptive filtering algorithms. He has also published works on speech enhancement, multi-channel microphone array and blind deconvolution algorithms. His other research interest includes education data mining, machine learning applied to education data. Andy currently serves as an Associate Editor in the IEEE Trans. Audio, Speech and Language Processing and the Journal of Multidimensional Systems and Signal Processing (Springer). He is the author/co-author of three papers awarded the "Best Student Paper Awards".

Yong Liang Guan (http://www.ntu.edu.sg/home/eylguan/) obtained his Ph.D. degree from the Imperial College of London, UK, and Bachelor of Engineering degree with first class honors from the National University of Singapore. He is a tenured associate professor at the School of Electrical and Electronic Engineering, Nanyang Technological University, Singapore, where he is now the Head of two industry collaboration labs: NTU-NXP Smart Mobility Lab, Schaeffler Hub for Advanced Research (SHARE) at NTU. His research interests broadly include coding and signal processing for communication systems, data storage systems and information security systems. He is an Associate Editor of the IEEE Transactions on Vehicular Technology.